New Politics, Progressive Policy

Series Editors: Sheila D. Collins, Emeritus Professor of Political Science, William Paterson University and Bradley Macdonald, Professor of Political Science, Colorado State University

New Politics, Progressive Policy, explores the changing configurations of world power and their implications for politics and policy in the twenty-first century. The series consists of politically engaged books that explore both the new challenges and progressive openings that are presented by the erosion of American hegemony and the jockeying for power among emerging economies, the positive and negative implications of non-state actors—social movements, NGOs, terrorists, global finance—and the challenges to global stability posed by climate change, global economic meltdown and inequality within and among nations. The series aims to provide scholars, students, researchers, policy makers, political activists, and the general public with a critical analysis of the political, economic and cultural developments generated by these changing configurations of power. These books may represent new theoretical approaches to important political/policy issues; comparative politics and policy studies; a new way of looking at a policy arena; emerging trends in political development and international relations; or revisionist readings of history. *New Politics, Progressive Policy* reflects the political vision of the Caucus for a New Political Science of the American Political Science Association, the largest and most progressive caucus within the association.

Other titles
The Global Sex Trade:
Economics, Policy and the State
Karie A. Gubbins

THE SCRAMBLE FOR AFRICAN OIL

Oppression, Corruption and War for Control of Africa's Natural Resources

Douglas A. Yates

PlutoPress
www.plutobooks.com

First published 2012 by Pluto Press
345 Archway Road, London N6 5AA

www.plutobooks.com

Distributed in the United States of America exclusively by
Palgrave Macmillan, a division of St. Martin's Press LLC,
175 Fifth Avenue, New York, NY 10010

British Library Cataloguing in Publication Data
A catalogue record for this book is available from the British Library

ISBN   978 0 7453 3046 4   Hardback
ISBN   978 0 7453 3045 7   Paperback

Library of Congress Cataloging in Publication Data applied for

10   9   8   7   6   5   4   3   2   1

Designed and produced for Pluto Press by Chase Publishing Services Ltd
Typeset from disk by Stanford DTP Services, Northampton, England
Simultaneously printed digitally by CPI Antony Rowe, Chippenham, UK and
Edwards Bros in the United States of America

Contents

List of Tables, Figures, Boxes and Maps

TABLES

FIGURES

BOXES

Abbreviations and Acronyms

ADI	*Acção Democrática Independent*
BEAC	Central Bank of Equatorial Africa
BPI	Bribe Payer's Index
BRP	*Bureau de Recherches de Pétrole*
CAR	Central African Republic
CNOC	Chinese National Oil Company
CNU	Cameroon National Union
CPI	Corruption Perception Index
EITI	Extractive Industries Transparency Initiative
ERHC	Environmental Remediation Holding Corporation
FAC	*Forces Armées Congolaises*
FIBA	*Banque Française Intercontinentale*
FLEC	Front for the Liberation of the Enclave of Cabinda
FNLA	National Liberation Front of Angola
GCB	Global Corruption Barometer
HDI	Human Development Index
IACC	International Anti-Corruption Conference
IMF	International Monetary Fund
JDA	Joint Development Agency
JDZ	Joint Development Zone
JNOP	Japanese National Oil Company
MDFM	*Movimento Democratico Forças da Mudança (São Tomé e Príncipe)*
MLSTP	*Movimento de Libertação de São Tomé e Príncipe*
MNC	multinational corporation
MNR	*Mouvement National de la Revolution*
MPLA	Popular Movement for the Liberation of Angola
NGO	non-governmental organization
ONGC	Indian Oil and National Gas Corporation
PCD	*Partido de Convergência Democraticá*
PCT	Congolese Labor Party
PDGE	*Partido Democratico de Guinea Ecuatorial*
OPEC	Organization of Petroleum-Exporting Countries
PDVSA	*Petróleos de Venezuela*
PWYP	Publish What You Pay
Sinopec	China Petroleum and Chemicals Company

STPetro	*Sociedade Nactional de Petróleo de São Tomé e Príncipe*
SOGARA	*Société Gabonaise de Raffinage*
Sonangol	*Sociedade Naçionao de Combustiveis*
SPAEF	*Société des Pétroles d'Afrique Equatoriale Française*
SPLA	Sudan People's Liberation Army
MLSTP	*Movimento de Libertação de São Tomé e Príncipe*
TI	Transparency International
UNIPEC	China International United Petroleum and Chemicals Company
UNITA	National Union for the Total Independence of Angola
UPC	*Union des Populations du Cameroun*
USCG	United States Geological Survey
WGI	World Governance Indicators (of the World Bank)

Acknowledgements

Over the past 20 years there has been a handful of professor/mentors including Edouard Bustin, David Gardinier, and Irene Gendzier, who have helped me to advance with my larger research agenda on African oil. Thanks also to John Berg I presented my research in Fukuoka, Japan, at the world congress of the International Political Science Association, where Georgy Kastiaficas invited me to propose my book project for the New Political Science series. Later on Sheila Collins took over, and sent the proposal on to Pluto Press' David Castle, who then involved Will Viney, Sian Mills, and Robert Webb, and Melanie Patrick on my project. Others who have had a hand either in the reviewing, editing, or directly/indirectly contributing to my long-term research for this book include: Tim Hughes, Andreas Mehler, Rudolf Traub-Merz, Matthias Basedau, Ian Gary, Adekeye Adebayo, Kaye Whiteman, Antoine Glaser, Brent Gregston, Geert van Vliet, Alain Beltran, Karen Klieman, and Kenneth Omeje. Finally, I would like to thank my parents, William and JoAnn, my brother James, and my wife Corentine, for their longtime support.

Introduction

The "oil curse" is a shorthand expression that denotes a series of dysfunctions—economic, political, governmental, and security—which are strongly associated with oil-dependency. Oil-dependent countries suffer from enclave industrialization, limited economic diversification, and vulnerability to price shocks, decay in their manufacturing and agricultural sectors, declining terms of trade, misguided economic policies, and a fundamental neglect of human capital. Economically these states have tended to neglect their human development because they are blinded by their resource wealth, which transforms them into oil-rentier economies. But in addition to these economic effects, African oil states also suffer from global patterns of domination and dependence, in some cases neocolonialism, in all cases multinational corporate exploitation, and well-intentioned but dangerous meddling in their governance by international organizations. Furthermore, they are postcolonial states, already fragile, that have been weakened by the corrupting influences of oil money, with their leaders reduced to kleptocrats, their civilian regimes transformed into brutal police states, and aggravating a regional tendency of military rule, their armed forces turned into praetorian cliques, personal despotisms and veritable reigns of terror. Beneath these aberrations of state power live poor and deprived societies that have been traumatized by five centuries of bloody exploitation, handicapped by low levels of education and health, primitive economies of accumulation, high rates of unemployment, limited capital, and few opportunities for advancement. Torn by violent conflicts based on ethnocentrism, unfair distribution, status frustration and internalized inferiority complexes, the people who live in these oil-rich countries are prone to rebellion, insurrection, and civil war.

Since all of these dysfunctions are interrelated and since they interact in complex ways, I have found it useful to break them down into separate discussions, and to arrange them in a logical order, reflected in the structure of the following ten chapters.

The methodology I have adopted for this present demonstration is that of the *case study*, which sets forth theoretical propositions that are then submitted to a close examination of a country to see if the theory correctly explains the case. Case study method offers a scientific reality test than can illustrate the validity of a theory by grounding it in empirical–historical facts. Moreover, where two plausible rival theories are applied to the same case, this method can compare their relative explanatory powers. A case study can never prove that a theory is true. But it can place two theories side by side and through contrast and comparison prove that one theory is *better* than another, that is, it explains a greater number of facts, or explains them more robustly.

Any theory that claims that oil wealth causes poverty is counter-intuitive, because we all know of countries which have gotten rich from their oil. Therefore if we are to test the proposition that oil wealth is a cause, and poverty an effect, then we cannot simply hold other factors constant. Context matters. The paradox of plenty that we find in Africa seems to be contextually grounded in African realities. This was how I came to discover that my initial question about why oil-rich African countries were still so poor would have to involve a thorough investigation into many other forces also operating on these countries, such as historical legacies that handicap the present possibilities of development, or international actors that dominate and subvert national ones, economic laws that govern political choices, and corrupt political cultures that pervert economic policies. The advantage of the case study method is that it allows for the inclusion of a multitude of factors that in their accumulation provide a landscape of context. Unlike the more prominent large-n quantitative comparisons, which examine only a limited number of factors over a large number of cases, but along the way eliminate the multitude of details and variations which make each country unique, case study method includes the totality of variables, factors, idiosyncrasies, and accidents that more often than not provide the complete explanation for complex social phenomena. Thick description of a case study rarely has to claim that one thing causes other "ceteris paribus."

The weakness of case study methodology is its external validity—that is, its ability to generalize from the case. Case studies can be used to make analytical generalizations from the case to a theory, but are not so helpful in making statistical generalizations to other cases. In order to compensate for this inherent limitation, I have employed the *comparative method* and selected similar cases for cross-analysis. Several chapters tabulate a small number of relevant factors that are present or absent in African oil states. These tables allow for simple generalizations to be made across the cases, as well as for the appreciation of important differences that may limit such generalization. In the language of comparative method these are called "closed universe" comparisons. While the oil curse has been observed in other regions of the developing world, the conclusions in this book are limited to my sample of African countries south of the Sahara. I leave it to my readers to reason by analogy to those other countries and/or regions of the world that are similarly cursed by oil.

Speculating about why oil-rich countries in Africa are so very poor, this book takes into consideration three levels of analysis. First, it fathoms the world in which African oil is bought and sold: the *international system* of nation-states; the source of markets and modern machines; the motive force of commoditization; the builder of drilling platforms, pipelines, and supertankers; the role of multinational corporations, of capitalism, and foreign armies—in short, the larger global context that takes a filthy sludge and turns it into "black gold." Next it explores the workings of government and institutions of sovereign power, the *state*, seeking clues about how oil is granted to foreigners, and how oil money is distributed and accumulated. Finally, it looks below the level of government, through the façade of a postcolonial state, peering behind the scenes into the *domestic political system*, the field of action that structures the formal institutions of government, those rules of the game that regulate the distribution of wealth and power, deciding who gets what, when, where, and how, and presenting to different actors with competing interests a battlefield for conflict and a forum for compromise.

The first chapter, "Foreign States and Trade Relations," will map out the geopolitical forces that shape the oil business in Africa (such as reserves, production, trade flows) and provide an overview of its phases of growth: from colonial spheres of influence, to national enterprises, to global trading flows with new patterns of conflict and cooperation. The current penetration of China may free oil producers from the legacy of Western domination, but it also perpetuates old archetypal patterns of African "scramble." With peak oil production predicted by 2025, world scarcity will increase, which could turn today's East–West trade competition into tomorrow's oil war. A case study of Gabon provides a way to tell this story of the modern "scramble" for African oil, from its colonial exploration by the French SPAEF, through neocolonial exploitation by Elf-Aquitaine, to the penetration of this French sphere of influence by American and now Asian actors, most notably China. Gabon has already passed its peak oil production, and is now looking for other partners and solutions. Its pro-China policy has enabled it to diversify its dependency, and to develop its non-oil minerals resources—changing patterns of French domination—but has not enabled it to escape from the pattern of domination itself.

The second chapter, "Multinational Corporations and Nationalization," shows how African oil is still largely dominated by foreign oil corporations, which rule over their oil concessions with quasi-sovereign power. This is no longer the case in Latin America or Asia, where national champions have come to control their own petroleum resources. Since African regimes have all nationalized their oil, the question is "why?" This chapter enumerates and systematically rejects a series of rival answers—geography, slavery, colonialism, neocolonialism, multinational strategy, and dictatorship—in order to arrive by argumentative structure at the theory of collaboration. Foreign domination of African oil enclaves is the result of extraversion and the collaboration of indigenous elites. Those who seek only to end neocolonialism must recognize that, if former colonial powers have been replaced by new foreign actors, the same old pattern of collaboration remains. A case study of the Cabinda Gulf Oil Corporation, which has been pumping oil in Angola since the days of Portuguese colonial rule, provides a striking example of poor corporate governance. While Angola did manage to nationalize its petroleum resources and establish a national oil company, Sonangol, Gulf (ChevronTexaco) continues to dominate in Cabinda, and new offshore reserves have been sold off as concessions to other foreign corporations, creating foreign-run archipelagos where Sonangol plays only a passive role.

The third chapter, "International Organization and Governance," looks at the non-governmental organizations working on the problem of the resource curse, which have advocated a series of international initiatives for transparency and good governance. Some have lobbied the international financial institutions such as the World Bank to formulate voluntary initiatives for good governance in the African oil sector. Others have called for mandatory disclosures and funded professional NGOs such as Publish What You Pay (PWYP) to investigate and prosecute corrupt practices in African states. Unfortunately their small membership, lack of representativeness, narrow range of issues, and dependency on foreign donors have all combined to reduce their effectiveness. Chad was the veritable test case of the "World Bank Model," which sought to create oil governance structures that could that ensure oil revenues would go toward poverty alleviation. This served as a model for the Extractive Industries Transparency Initiative (EITI) to promote revenues transparency and improve macro-economic management of oil revenues. The Chad–Cameroon pipeline became the single largest foreign direct investment in sub-Saharan Africa. At first, Chad tried to use its oil revenues for military expenditures, with the World Bank freezing its accounts. But as hundreds of millions of dollars' worth of oil revenues came into the state coffers, Chad paid off its debt to the World Bank, and no longer adheres to the model.

The fourth chapter, "Rentier States and Kleptocracy," provides a synthetic discussion of the theory of the rentier state, i.e. one dependent on external rents for the bulk of its revenues. Financial autonomy from its domestic society reduces the democratic accountability of a rentier state, and funds corrupt patrimonial systems of rule. Its bureaucracy transforms into a rentier class, with a rentier mentality that isolates both position and reward from either hard work or merit. Spending unearned oil revenues to buy popular consent and purchase prestige symbols of development, its rentier economy suffers from a series of deep macro-economic distortions that worsen as oil revenues increase. Instead of attending to the needs of ordinary people, a rentier class is created that devotes the greater part of its resources to conspicuous consumption and jealously guarding the status quo. The longer this class remains in power, the greater its tendency to become a kleptocracy, where public revenues become the private wealth of ruler, clients, and kin. The case of Equatorial Guinea, which has enjoyed one of the highest rates of economic growth in Africa thanks to oil exports representing over 90 percent of public finance, illustrates how oil wealth, which should have been a blessing, has in fact been a curse. Not only have oil revenues funded one of the most brutal regimes in the region, prolonged the rule of a bloodthirsty dictator indefinitely, and lavished expenditures on presidential palaces and overseas estates, but the one-time opportunity to invest in public education, health, housing, and utilities has been lost. Oil production is already declining, and the revenues it has generated have largely been stolen by a rentier class, the Mongomo clan.

The fifth chapter, "Praetorian Regimes and Terror," undertakes an analysis of military rule. Professional soldiers rule all but a few of African oil states, coming to power through coup d'état or civil war. Despite removing their military uniforms and donning the costume of civilian presidents, these men came to power by use of force, and rule through the use of violence and terror. It is our oil money that makes their terror possible. Not only have they spent vast amounts of oil-rent on building armies and secret police, but when their subjects are unfortunate enough to live above the oilfields, they oppress them with unspeakable terror. Refusing to attend to the socio-economic needs of their people, they concentrate instead on remaining in power. The case of Congo-Brazzaville provides an example of oil-funded, "civilianized" military rule. President Sassou-Nguesso is a professional solider who has the dubious distinction of twice coming to power through the use of force. First he seized power in a 1979 coup d'état and ruled a Marxist dictatorship funded by oil revenues from the French Elf-Aquitaine. Then, after losing in the first and only free elections in 1992, he returned to power a second time by overthrowing the elected government in a bloody civil war in 1997. He and his family have amassed vast fortunes abroad, while his subjects live in abject poverty. But what is worst about his regime is that, in order to fund a lavish lifestyle, he has borrowed billions of dollars, using future oil production as collateral, stealing money in advance with these oil-backed loans, and so he has cursed future generations with a debt burden they will not be able to repay.

The sixth chapter, "Journalists and Intellectuals," provides an analysis of intellectual dissidence. Much of what we know about the abuses of oil and power in Africa comes from the courageous efforts by investigative journalists and intellectuals. Despite the relative weakness of the press, there is a tradition of critical writing in Africa that goes back to the struggle for independence. This tradition serves the democratic watchdog function of the mass media. But when government repression of the media is too strong, this watchdog role has been assumed by African intellectuals, who speak truth to power, and provide a common voice for people to mobilize in the absence of effective opposition political parties. The late Mongo Beti was one of Cameroon's celebrated intellectuals, using his fame as a novelist to criticize both the endemic corruption of the Ahidjo regime and collaboration with the French. Despite being banned in his native land, and being marginalized by the Paris-dominated techno-structure of *Francophonie*, Beti's works enjoyed international fame and could not be dismissed as easily as Anglophone critics from Cameroon's western region. Beti used his pen to shape public opinion, and, after his death, his words still influence ordinary people, and provide a common ground for those who oppose the regime of this oil-transit state.

The seventh chapter, "Political Parties and Elections," explores the possibility of opposition parties to bring about change through democratic elections. In a military regime, opposition parties are weak or non-existent and their calls for justice go largely unheeded. In a civilian oil-dictatorship the structure of public finances concentrates wealth and power in the hands of a rentier class. Since allocation of oil revenues is monopolized by the ruling elite, political opposition tends to focus its attention on how those revenues are allocated. Their solution for maneuvering for personal advantage within the existing setup tends to be superior to seeking an alliance with others in similar condition. But in a democratic state it is possible for opposition parties to campaign for changes in both leadership and public policy. Two African oil-exporting countries—Nigeria and São Tomé—are democratic, and in both cases, this is precisely what happened. The case of São Tomé offers a unique opportunity to test the proposition that democracy can resist the oil curse. President Fradique de Menezes is a businessman who broke away from the ruling party to found his own opposition party, Force for Change Democratic Movement [sic]. Oil discoveries brought foreign companies to the island archipelago, which had corrupted the former ruling party with signature bonuses for bids on oil concessions. Against this corruption Menezes launched a successful campaign in the 2001 elections. Once in power, he and his party launched new rounds of open bidding. Since then, they have formulated progressive programs designed to use future oil revenues for economic development and poverty alleviation of the people.

The eighth chapter, "Armed Struggle for Independence," theorizes about the relationship between oil and conflict in Africa, a veritable sub-field of international relations. Two rival schools of thought have become influential. The first is that rebellions are caused by "grievances" of marginalized peoples. The second is that rebels are motivated more by their "greed" for oil revenues. Rather than taking sides and perpetuating dispute, it is better to synthesize these two approaches. Both are factors that cause and fuel conflict. Since oil revenues tend to fund the state, most rebellions find themselves out-resourced by their governments. Since world opinion is shaped by our need to consume African oil, we tend to advocate peace without justice, and to ignore the original grievances behind the conflicts. The northern media tend to see these oil rebellions only as negative because they are violent. But they can effectuate positive change. The late John Garang was the most famous rebel in the long civil war between the north and south of Sudan. Centuries of oppression by northerners, followed by the construction of a southern identity under British colonialism resulted in a demand for self-rule by the south. When the north refused a referendum on independence, the Sudan People's Liberation Army (SPLA) arose to fight for it. This conflict was later fuelled by the discovery of oil in South Sudan. Waging a decades-long armed struggle, Garang eventually succeeded in negotiating a peace with both power- and revenue-sharing, a success that has inspired other rebellions, such as the Justice and Equality Movement in Darfur.

The ninth chapter, "Popular Resistance and People Power," takes a radical perspective on the organized, collective, and sustained efforts to promote social change that occur outside conventional politics, drawing support from marginalized groups. Such popular forms of civilian challenge to oppression and injustice, when they depend on non-violent action, are called "people power movements," and when they are suppressed, either disappear, or turn into armed insurrection. People-power movements are diverse networks united by a strong oppositional consciousness and characterized by diffuse leadership and democratic structures. They have spontaneously appeared in most African oil-rentier states. Some are part of the emerging anti-globalization movement and have environmental concerns. While oil consumers in the global north have treated them as problems, because they can shut down oil production and cause higher gasoline prices, we should consider them instead as solutions for those very same oppressed Africans who we want to help with our transparency and good governance initiatives. Social activism by the late Fela Kuti, the most famous musical dissident in Nigeria, and the late Ken Saro-Wiwa, the most famous popular leader of the Ogoni people of the Niger Delta, railed against the abuses of the Nigerian military regime, and the negligent oil pollution by multinational oil companies such as Shell, which resulted in spontaneous people-power movements that both mobilized the oppressed masses and attracted international attention to the problems of their country: a change in consciousness. Saro-Wiwa's execution by the Abacha regime only turned him into a martyr and helped, after the transition to civilian rule, to change government and corporate practices toward revenue-sharing with the people of the Niger Delta.

The tenth chapter, "Unscrambling the Scramble for African Oil," offers five different solutions to the problem of the oil curse. First, it outlines a menu of policies designed to control corruption, based on the new-institutionalism approach, which focuses on changing the selection of agents, changing the system of rewards and punishments, gathering information of the analysis of risk of corruption, and restructuring the relations between agents and users to reduce the monopoly and discretionary power of agents. Second, it explores a radical proposal to directly distribute oil revenues to the people, which would bypass the corrupt regimes and re-enfranchise African citizens. Despite sounding like a Utopian dream, this proposal of direct distribution of oil revenues has in fact been implemented in the state of Alaska, and poses fewer problems in implementation than the present international transparency and good governance initiatives. Third, it reviews the Venezuelan model of investing in social development as implemented by Hugo Chavez's Bolivarian Revolution. Harnessing oil revenues for human development, this model makes use of imaginative new social programs known as "missions" that provide education, health services, housing, and job creation for Venezuela's urban and rural poor. Fourth, the solution of a boycott of African oil is suggested, including the possibility of using "smart boycotts," which use modern internet hedge funds to attack oil corporations, and the creation of a "blood oil" campaign based on the successful "blood diamonds" campaign, which instituted the Kimberly process. Finally, the solution of reducing, and when possible stopping, our own consumption of oil is advocated, requiring a fundamental change in consciousness. This solution would not only eliminate the oil curse in Africa, but would also address the serious environmental problems for which current consumption patterns are responsible.

The primary aim of this book is to explain the terrible paradox that contrasts the abundance of oil wealth in certain African countries with the poverty of ordinary people who inhabit them. When I first set myself to solve this problem more than 20 years ago, I thought the solution could be propounded very briefly, but I soon found that to render it probable or even intelligible it was necessary to discuss certain more general questions, some of which had hardly been breached before. The discussion of these related topics has occupied more and more space, and the enquiry has branched out in more and more directions, until that single question has expanded into ten. Whether the explanation I have offered of the paradox is correct or not must be left to the future to determine. I shall always be ready to abandon it if a better can be suggested. Meanwhile, in committing the theory of the rentier state in its new form to the judgment of the public, I should like to guard against a misapprehension of its scope that appears to be still rife. If in this present work I have dwelt at some length on the effects of oil rent, it is not, I trust, because I exaggerate its importance in the history of

Africa, still less because I would deduce from it a whole system of explanation for the continent's problems. It is simply because I could not ignore the subject in attempting to explain the significance of an economic sector on the society, politics, security, and government of oil-dependent countries. But I am so far from regarding oil rent as of supreme importance for Africa, that I consider it to have been altogether subordinate to other factors, and in particular to the lust for wealth and power, which, on the whole, I believe to have been probably the most powerful force in the making of African history.

Part I

Power from Above

1
Foreign States and Trade Relations

COLONIALISM, NEOCOLONIALISM, AND GLOBALIZATION

For half a century, foreign oilmen followed a *colonial pattern* of investment in Africa. Money came from Europe. Geologists came from Europe. Companies came from Europe. They tended to go where they spoke the official language and enjoyed citizenship rights. They represented imperial interests in closed political economies. Each worked within their own respective "sphere of influence." Colonial powers preferred to sell concessions to their own companies to pioneer petroleum production in a context of legal rights and privileges. British oil companies ruled over Nigerian oil because Britain ruled over Nigeria, and had the capital and mining technology to do so. French oil companies did the same in French Africa. Spanish oil companies also tried this offshore in Spanish Guinea, unsuccessfully. But the Portuguese were too poor and technologically weak to develop oil in Angola, so they sold concessions to the Americans. In all cases, colonial oil was theirs to sell. The history of this oil adventure is well known.

In British Nigeria, D'Arcy Exploration (later BP) first started reconnaissance work in 1937. The Second World War interrupted their drilling. Then, in 1951, Shell Oil joined d'Arcy and invested millions of pounds sterling in Nigerian exploration, finally producing oil in 1957. (Howarth 2007) BP was kicked out in 1978, but Shell remained in the Niger Delta.

In French Africa, the Société des Pétroles d'Afrique Équatoriale Française (SPAEF), a colonial subsidiary of the Bureau de Recherches de Pétrole (BRP), conducted the earliest seismic tests and shallow-water exploration. The BRP was a state organism created by the French government to develop colonial oil resources in North Africa. After the Algerian war of independence, when France lost its oil to the Arab nationalists, this state machinery and personnel migrated down into Equatorial Africa, where it founded local subsidiaries that adopted the host country names: Elf-Gabon first exported oil in 1957, then Elf-Congo in 1969, then Elf-Cameroon in 1973. These subsidiaries were run by the Paris-based mother company, Société Nationale Elf-Aquitaine, and continued to dominate the old sphere of influence, helped by a shadowy network of soldiers and spies under the infamous Jacques Foccart. Elf usually resisted efforts by American majors to penetrate its African oil zone, so when it invited foreign partners to share in the risks of investment, Elf always remained the principal operator. French oilmen were pioneers of the *neo-colonial pattern* of petroleum investments. (Yates 1996, 2006)

In Angola, however, the Portuguese oil companies were not sufficiently advanced, nor were their domestic markets large enough to justify enormous colonial investments. So Lisbon invited Sinclair Oil, an American firm, to conduct exploratory drilling in the Cabinda Enclave during the interwar years (1918–32). When it left during the Great Depression, Lisbon invited another American oil company, Gulf Oil, in 1957 to invest in its Angolan possession. So the Americans began exporting oil from Cabinda in 1968, and during the long Angolan war of independence, Portuguese soldiers protected them from Cabindan separatists. Even after independence, the new Angolan regime did the same. The MPLA founded a state oil company, Sonangol, to be its sole legal concession holder. But it needed foreign capital, technology, and markets to sell its oil. So the regime had no choice but to farm out the enclave, leaving Gulf Oil (Chevron) with two-thirds of production. (Minter 1972)

In Spanish Guinea, the national oil company, Repsol, tried but failed to find oil in Spain's only African possession. After independence in 1968 they continued to explore offshore until the Equatoguinean regime broke with Spain and the West to become a client of the Soviets. Few investments were made by the majors during this period. But when the first ruler was overthrown by his nephew in a 1979 military coup, the new regime offered its offshore waters to American investments. Mobil Oil thus entered into this hispanophone sphere of influence, followed by a legion of American oil independents, who transformed Equatorial Guinea into an important African oil-exporting country. (Liniger-Goumaz 2005)

Old colonial patterns of investment can be understood as an effort by the European oil companies to monopolize petroleum resources in their respective spheres of influence. These patterns became *neocolonial* when African countries were no longer colonies, yet they still depended economically and militarily on foreign powers. By the end of the twentieth century, four foreign majors dominated Africa: Shell, TotalFinaElf, ExxonMobil, ChevronTexaco. They produced around 75 percent of all oil exported from the region. (Copinschi 2001: 34) Two were European, and two were American. The fall of British, French, Portuguese, and Spanish monopolies had been paralleled by the rise of American capitalism. American majors in the European spheres of influence may have taken longer to arrive in those oil enclaves. But their arrival was a by-product of America's rise to global power during the Cold War. During this period of African history, superpower rivalry determined by ideology replaced the older patterns of collaboration. The West was often on the wrong side, choosing men who would collaborate with multinational corporate designs for Africa and assassinating genuine nationalists and pan-Africanists who might, had they lived, taken their countries in a different direction.

The ex-colonial European powers, by virtue of the enormous and in many ways profoundly formative impact of their colonial policies on patterns of African economic growth, international trade, state formation, recruitment of indigenous leadership, and linguistic and religious development, were from any historical perspective far more important non-African actors than the two recently arrived superpowers. At the same time, the ex-colonial powers were clearly a part of the West, which found itself pitted against the East. In this respect the presence of Western Europeans in Africa—whether as technicians, soldiers, teachers, policy advisors to governments, or private entrepreneurs—meant that Western leverage on Africa was considerably greater than that of the Eastern bloc. While the United States worried about Cuban troops in Africa, the Soviet Union worried about French troops in Africa and the potential they and their allies had for retaining the continent as a largely European sphere of influence.

A special complication during this period of intense superpower rivalry was that each superpower, anxious to contain its rival but equally anxious to avoid a direct confrontation in which it might not be able to prevail over that rival, sought to exert international influence indirectly through client states, which it supported in the expectation that they would independently advance the superpower's interests. Each superpower further assumed that its rival acted in the same indirect fashion and exercised ultimate control over the foreign policy of client states, which served as mere proxies or surrogates for their powerful patron. The US government perceived Cuban military activity in Angola as the not-terribly-subtle means by which Soviet power was extended in southern Africa. In quite a similar way, the Soviet Union perceived France, Belgium, South Africa, Morocco, and Mobutu's Zaire as surrogates for the African interests of the capitalist hegemon, the United States. The oil industry in Africa was shaped by and for triumphant Western capitalism. But after the Cold War, other foreign actors from Asia soon followed into the breach the Americans had opened up. This new "Scramble for African Oil" is a *pattern of globalization* that has generalized foreign oil investments, which had previously been exclusive domains of former colonial powers. The arrival of Japan, Malaysia, India, and China into these African enclaves long dominated by the West has been the most striking change caused by this pattern of globalization. The first to invest was the Japanese National Oil Company (JNOC), which entered Zaire in the 1970s. Japan was totally dependent on oil imports, and was an American ally during the Cold War. In the 1990s the Malaysian national oil company, Petronas, and the Indian Oil and Natural Gas (ONGC) used their technology and finances to scramble for oil in the interior of Sudan.

The Chinese National Oil Company (CNOC) and Sinopec also arrived in Sudan in 1995. What is most startling about China's entry was the speed with which it penetrated traditional spheres of influence. Although China had *purchased* Angolan crude from the Marxist regime as early as 1988, it was not until it joined an oil consortium and started *producing* crude oil in Sudan (1999) that the world first took notice. Buying exports from Congo (2000), opening offices in Libya to explore the desert (2001), sending exploration crews into Nigeria (2002), winning projects in Algeria (2002), buying oil cargoes from Equatorial Guinea (2002), signing agreements to import from Gabon (2004), beginning trial production in Darfur (2004), signing contracts in Mauritania (2004) and production-sharing deals with São Tomé (2005), sending seismic crews to Ethiopia (2005). All of a sudden China was in every oil country, raising new questions: Would Chinese oil companies change investment patterns in Africa? And would their strategy provoke an oil war with the Americans? (Le Pere 2006)

RESERVES, PRODUCTION, PEAK OIL, AND WAR

Before answering these questions raised by Chinese entry into the scramble, it is important to explain the basic facts of African oil. Our sources for these facts are the oil companies, oil-exporting governments, and oil business analysts. There are no independent sources of data, although international organizations give that

China is the world's second-largest consumer of oil, and a quarter of its imports come from Africa

Map 1.1   China in African oil since 1988

appearance by publishing data actually derived from these primary sources. Even we free-thinking scholars depend on oil business sources. This raises problems of reliability. The oil industry is not transparent. Oil-rich states are often corrupt. And oil analysts frequently compete with rival data, rival theories, and rival methods. Add to this the different systems of measurement (barrels of oil, tonnes, c.i.f., f.o.b., proven versus estimated reserves) and you see the problem.

Africa south of the Sahara is estimated to possess around 5 percent of world oil reserves, according to British Petroleum. It may have over 53 billion barrels in total proven offshore reserves. West Africa is viewed by the industry as the world's leading offshore oil region. Business analysts largely agree that the Atlantic trend is an oil-industry Eldorado. More capital expenditure has been invested in the Gulf of Guinea than in any other offshore oil region. But if we look beyond such aggregate figures of the whole African oil sector, we see a picture of uneven growth. Some countries are discovering new reserves. Others are exhausting theirs. Some countries are increasing production. Others are declining. Some countries are trading with the West; others are going to the East.

British Petroleum publishes a statistical review of energy (available free on their website). On average, oil production in Africa rose 12.4 percent between 2007 and 2008, but a quick glance at the changes by country of origin show how different the picture is between those countries where output is *increasing*, such as Angola (+9.1 percent), Sudan (+2.6 percent), Congo (+12.3 percent), and Gabon (+2.2 percent), and countries where production is *declining*, such as Nigeria (–8 percent), Chad (–11.5 percent), and Equatorial Guinea (–2.1 percent). These variations in oil output have produced important change in the rankings. In the past, Nigeria was the number-one oil producer in sub-Saharan Africa (North Africa is treated separately by convention). Since 2008, Angola has been number one. These changes are explained by huge capital expenditures made in deepwater oilfields off Angola, and in onshore southern Sudan. They are also explained by the ongoing conflict in the Niger Delta. But the declines in Chad and Equatorial Guinea reflect a different "structural" reality: proved reserves are declining. Chad pumps oil from a handful of fields in Doba, but without new fields brought on line, its production will decline. Equatorial Guinea is depleting its largest reservoirs, and the oil companies are using horizontal drilling and/or sub-surface injection to squeeze out the last drops. Equatorial Guinea does have more oil offshore, but unless those reserves are proven (i.e. drilled) and brought on line, its production figures will decline. Small producers such as Cameroon, Democratic Republic of Congo, or Mauritania (tabulated by BP as "Other Africa") represent less than 1 percent of total African output. All of this makes a very simple point: oil is a scarce and non-renewable resource. Once "peak" production is reached, African reserves will fall ineluctably towards exhaustion.

According to BP, proved reserves in sub-Saharan Africa range from: 36.2 billion barrels (Nigeria), to 13.5 billion (Angola), 6.7 billion (Sudan), 1.7 billion (Gabon), 1.9 billion (Congo), 1.7 billion (Equatorial Guinea), 0.9 billion (Chad), and 0.6 billion ('Other Africa'). These "proved" reserves are limited to those reservoirs that geological and engineering information indicates with reasonable certainty can be recovered in the future under existing economic and geological conditions. As the price of a barrel of oil increases, so do the proved reserves. As new technological advances drill deeper and pump better, more oil reserves will be proved. One paradoxical phenomenon is that, as oil production has increased, proved crude reserves have also increased, because of the very act of drilling itself (i.e. proof). BP's publication of proved reserves is not a conspiracy by the oil company to create in us a false sense of scarcity; rather it reflects the limits of forecasting the future, and the desire to base estimates on hard evidence rather than on geological speculation.

The China question is related to peak oil. So long as Africa is in an oil boom, with new discoveries and rising production, Europe, America, and Asia can scramble with impunity. But future conflict or cooperation depends ultimately on the timing of peak oil and the size of proved reserves. BP publishes a "reserves-to-production" ratio—R:P—that represents the length of time that remaining reserves would last if oil production were to continue at the previous year's rate. It divides remaining reserves at the end of a year by that year's oil production (R/P = reserves/production.) This statistic includes unproved reserves that are known geologically but not yet proved by drilling. According to BP reserve–production ratios, the future will vary from: 45.6 years (Nigeria) 38.1 years (Sudan), 37 years (Gabon), 21.3 years (Congo), 19.7 years (Angola), 19.4 years (Chad), 12.9 years (Equatorial Guinea), and 12 years for the rest of Africa. This ratio, while useful, does not really answer the question of peak oil, for it assumes that production will be the same in the future, and that no new discoveries will be made. But these are variables, not certainties. The R:P ratio cannot say with certitude exactly when oil production will peak. It can, however, say that some countries will peak before others.

An alternative approach is the "Hubbert method," named after Marion King Hubbert, a Shell geophysicist in Houston, who published controversial calculations in 1956 showing that US oil production was going to peak in 1970 and thereafter rapidly decline. At the time, the US was the largest oil producer in the world, and few people took Hubbert's idea seriously. But he was proved right. American production did peak at 9 million barrels per day in 1970, and it has since then declined to 6 million barrels a day. The exactness of Hubbert's prediction led oil companies to use his method to estimate the capacity of their productive oil fields. Hubbert had data only for the United States, but in the late 1990s geologists used world oil data to evaluate world reserves. According to their calculations the "Hubbert peak" at the world scale would arrive in the first decade of this century. That is, world production has already peaked! This startling prediction resulted in the publication of several best-sellers, including: *Hubbert's Peak: The Impending World Oil Shortage* (2001) by Kenneth Deffeyes, a colleague of Hubbert at Shell who now teaches geo-science at Princeton University, and *Out of Gas: The End of the Age of Oil* (2004) by David Goodstein, a physicist at California Institute of Technology. You may have heard about "peak oil" when these books came out.

Hubbert used three methods. First, he borrowed a logic used by population biologists, who had observed that whenever a new species develops in an area possessing abundant resources, its population growth becomes exponential. But when the population becomes large enough so that the resource no longer seems unlimited, its growth slows down. This phenomenon is the same in oil prospecting. As new oil discoveries get smaller, so do the size of oil reserves. Hubbert showed that the growth of US oil discoveries had been in decline since the 1950s, and that by extrapolation from past data, it was possible to determine when growth in the size of reserves would end. At that instant, he argued, all of the oil that was under the earth would have been discovered. More important, the total quantity of oil that ever existed would be equal to the quantity already extracted plus the reserves known to still be under the earth. (Total oil = total extracted oil + total oil reserves) The second method used by Hubbert assumed that, over the long term, when you trace the curve representing oil discoveries already made, you obtain a "bell-shaped" curve—that is, a curve that starts by increasing to reach a summit that is never surpassed, and then falls down towards zero. Using world data, total world oil is estimated at around 2 trillion barrels, around half of which has already been extracted. According to Hubbert, when this half-way point is reached, we will have reached the 'peak' (or summit) of the oil production curve. (Total extracted oil = total oil reserves.) His third method makes the following observation: The curve of total oil extracted at any given moment in time is parallel to the curve of oil discoveries, with a delay of a few decades. In other words, we pump oil around the same rhythm that we discover it, but several decades after it is discovered. (Past discovery curve = future production curve.) Therefore he reasoned that the rhythm of discoveries could serve to predict the rhythm of extraction in the future. In the United States this is exactly what happened. On the world scale, discoveries started declining several decades ago, in both number and size, and therefore the peak for oil discovery has already occurred. Now we are waiting for the peak in production to follow.

Of course, not all geologists agree with his methods or his assumptions. The United States Geological Survey conducted a study in 1995–2000 where it concluded that, if there was a 95 percent probability that world oil reserves are 2 trillion barrels, there is also a 50 percent probability that there are actually 2.7 trillion barrels. The USCG challenged Hubbert's first assumption that new discoveries will continue to decline, and included their own assumption that as oil prices increase, the quantity of recoverable crude will increase to make up this difference. This additional 0.7 trillion barrels could delay the Hubbert peak by around ten years. (Goodstein 2004: 26)

Let us assume, for purposes of answering our question, that BP reserves:production ratios are correct. If the total proved oil reserves for North and sub-Saharan Africa are estimated at 117.5 billion barrels (or 9.5 percent of the world total), then with an average production of 10.3 million barrels a day, the reserves:production ratio for all of Africa is 31 years. Now in our region of interest, sub-Saharan Africa, only Nigeria and Sudan will still be producing oil by 2030. What are the chances that the US and China will go to war over these two countries, i.e. fight over ten more years of African production? Now assume that production increases, and also that oil prices continue to rise, which makes expensive oilfields more economically viable, and thus increases the proved-reserves figures. Assume that new offshore technologies are introduced, resulting in production of reserves that are still unproved but geologically known. Or assume that world-class oil discoveries are made in the interior (in the Congo basin, for instance, where the oilmen have always speculated that a huge field exists somewhere in the Jurassic). Do the chances of an oil war increase?

This kind of reasoning is based on a theoretical assumption: countries go to war over oil. We shall discuss this theory in Chapter 8. For now it is enough to recognize that our question of whether China and the United States will go to war over African oil presupposes that major powers go to war over resources. This theory has empirical foundations in the history of war, but it is far from being a perfect predictor of conflict. In the end, the question depends less on African oil, or even on world oil reserves, and more on the causes of war and peace. Theories of conflict in international relations have debated the causes of war for more than two thousand years. Resource conflict theory does not stand alone.

A rival theory, nuclear deterrence, claims that nuclear powers don't go to war with each other. Empirical evidence supporting this is the historical fact that nuclear powers have not gone to war with one another. One reason is that the costs of nuclear war far outweigh any benefits. According to deterrence theorists, nuclear war is unwinnable. There is no more ironclad law in international relations theory than this. Nuclear states do not fight wars with each other (major premise). The US and China are both nuclear powers (minor premise). Therefore they will not go to war (QED). But this "law" has one counterexample. India and Pakistan are nuclear powers. After Pakistan acquired the bomb, deterrence theorists confidently believed that India would not risk going to war, despite a long history of bloody conflict. Then in 1999, one year after an exchange of nuclear tests, they did fight a war in the mountains along the line of control separating the portions of Kashmir controlled by each country, near the Indian town of Kargil. You may object that this was a border war between two enemy states—it was not a nuclear war, and not an oil war. But the Kargil affair is disturbing because it demonstrates that nuclear-armed states can fights wars.

For those readers who are not accustomed to reasoning with international relations theory, it is very important to understand that these are theories, not really "laws." Both the theory of oil war and the theory of nuclear deterrence are *inductive* inferences (i.e. generalizations based on particular facts). Therefore you shouldn't make *deductive* inferences from them, as you would from scientific laws. This is why the logical syllogism just given above, which treats the assumption that nuclear states will not go to war as an axiomatic truth, is bad reasoning. Such theories are predictive statements about the relative probability of war—you must decide which one is more robust. See them as rival predictions about alternative possible futures, where human choice plays a major role. Thucydides, father of international relations, told us how Athenian leaders made bad choices which led them into a fratricidal Peloponnesian War. But they had choices.

To the extent that oil is a cause of conflict, and that nuclear states can and will fight wars, then we should seek to remove that potential cause of nuclear conflict. This can be done by reducing and, when possible, eliminating our own oil dependency. It would be unwise to naively saunter down the primrose path to nuclear destruction on the false belief that nuclear deterrence is 100 percent successful. It would be equally unwise to store canned food in our cellars fatalistically awaiting imminent nuclear holocaust. What can be used by states, in a world with no common power, is the timeless tool of diplomacy. So far the governments of the United States and China have controlled the potential for oil conflict preconditioned by the scramble for African resources. We can try to influence our governments. But even in democracies foreign policy is seldom made by ordinary citizens. Our economies are oil dependent, and our leaders are not free to ignore the national interests. But we do not need to wait for government leaders to stop our oil consumption for us. What we need is to reduce our own oil consumption. For this we don't need leaders, or repressive laws to coerce us to act in this way or that. Nor do we need to fight wars over oil reserves. What we really need is a transformation of our consciousness, and the will to act responsibly as individuals in concert with others. In world politics such relations are called "transnational." They are distinguished from "international" relations, because they concern the behavior of individual citizens rather than their states.

CASE STUDY: NEOCOLONIALISM IN GABON

A scientific case study is not simply the selection of a case. It is a case *of* something. A case study is a case of a well-defined theory. Gabon has been selected to represent a case of *neocolonialism*. This choice was made because most of us already understand that colonialism was a form of foreign domination, and therefore the domination of a colonial oil enclave by the colonial power was merely a logical extension of the overall imperial structure. Also most of us understand that in our age of globalization, old colonial spheres of influence have been penetrated by new actors from around the world. But what most of us might not understand is how dominance–dependence relationships between former colonies and their colonizers were able to survive after decolonization. Gabon is an oil-rich enclave of enduring French influence that has been dominated by French companies long after independence in 1960—it is not the only case of French neo-colonialism, but it is an extreme case, and therefore highlights the problem with illuminating intensity.

Neocolonialism was first postulated by the founding father of Ghana, Kwame Nkrumah. It was born out of his disillusionment with the failure of Africans to start a new history after achieving independence. That the colonial past was of great significance in determining the future of post-colonial states was recognized by all,

Map 1.2 Gabon

even as the process of decolonization proceeded. But colonialism, by globalizing the European model of the sovereign nation-state, determined that the primary object of anti-colonial nationalism would be the transformation of colonies into independent nation-states. Nkrumah believed that such political independence must also bring economic independence, cultural independence, and an escape from colonial mentalities imposed by five centuries of European domination. The struggle for freedom did not end with political independence, but continued as a struggle for economic, strategic, and cultural freedom from all legacies of colonial rule. He renamed the Gold Coast Ghana, nationalized its British colonial industries, and introduced African socialism. If his socialist experiment was a failure, his theory of neocolonialism was not.

There are three preconditions for neocolonialism to exist. First, a country must have been colonized by another. Second, it must achieve formal independence from that colonial power. Third, a dominance–dependence relationship must persist between them after decolonization. All three conditions are met in the case of

Gabon, which was colonized by France, granted independence, and thereafter remained in a dominance–dependence relationship with its former colonial power. It is best to examine these three phases in chronological order, because neocolonialism is a "path dependent" outcome where past choices made along the way significantly influenced the future destination.

Before the arrival of Europeans, there was no such thing as 'Gabon.' There was a place, and people who lived in that place, but it was not territorially demarcated in the shape we see on the map today, and its people were certainly not called Gabonese. That name came from the Portuguese explorers who discovered the estuary in 1472. It was "shrouded" in a fog so thick that they gave it the name *gabão*, meaning "hood" or "sleeve." The Bantu peoples who they met in this estuary called themselves by many different names: Mpongwe, Orungu, Vili. They were not sovereign "nation-states" in the European sense—they were African societies. Their basic units of political organization were the extended family, the lineage, and the clan. (Only on the Loango coast were institutions then developing on a larger scale into kingdoms.) In 1843 France founded a post in this "Gabon Estuary" to promote commerce and combat the slave trade. It secured recognition of its sovereignty through treaties with the clan heads as the first step in excluding rival claims by the British, American, and German traders. Over the next four decades treaties were signed with all the coastal tribes, and explorers pushed into the thickly forested interior. Famous adventurers in this chapter of the scramble for Africa were Paul de Chaillu and Savorgnan de Brazza. Gabon's territory was defined not by the Africans, but by the European colonizers, through a series of French treaties with colonial Belgium (1885), Germany (1894), and Spain (1900). It was not until 1918 that Gabon's territory was given its modern shape: Gabon is a geographical construct of France.

At first Gabon was administered by French naval officers, but after the Berlin Conference (1884–85) it began to acquire a more typically colonialist character. France instituted a hut tax and required unpaid labour from those who could not pay it. During the years of exploration and occupation of the interior France required villages to provide food for forest outposts, and committed its worst abuses during the phase called the "concessionary regime" (1898–99), which denied all land rights to the indigenous people. Land belonged to the state, not to the people who lived on it, and it was sold to French concessionaires. These were colonial firms who emptied the countryside of all available resources, backed up when necessary by French-run Senegalese militiamen. Forced labour caused people to flee, and to abandon their farms, which resulted in famines and epidemics. Concessionary companies nearly destroyed the existing commercial networks and sent the economy into a slump from which it would not recover until large-scale forestry began after the First World War. The official colonial policy was *assimilation*: the creation of French-speaking African elites who secured appointments as chiefs or as representatives in consultative bodies established by the Popular Front (1936–37). They learned French customs and adopted French styles of dress, read French books and papers, and learnt French ideas. Much of the post-war political leadership would come from these elements of Gabonese society. The two world wars generated trends of decolonization throughout the colonial empire, which France tried to contain by promoting advancement of blacks in its African territories within a structure dominated by Frenchmen. After the war, the French Fourth Republic (1944–58) put an end to the worst abuses of the old concessionary regime, and granted certain basic rights. (Gardinier and Yates 2006: *lii*)

By then the economy was dominated by French foresters, who had organized themselves politically as a powerful lobby to protect their economic interests. In those days Gabon was a sleepy timber enclave. It was the leading African producer of tropical hardwoods, with timber making up 80 percent of the colony's exports at independence. Yet the foresters had not modernized the sector, which essentially exported logs to France. Large trees were felled near the rivers, and then simply floated downstream to the ports, to be shipped off raw and unprocessed as timber. Mahogany, ebony, and walnut were exploited in large quantities, the most important species being okoumé, a type of soft mahogany used in plywood production. Trade in this species was so lucrative that the French called champagne "okoumé juice." When independence came to Gabon these foresters needed a willing pro-French collaborator who would protect their business interests. The man they chose was a Fang politician, Léon Mba, the mayor of the colonial capital Libreville. The Fang were the largest ethnic group, and he worked for a forestry firm. With their backing, his party won the first legislative elections ever held (1957) and he became the first prime minister of Gabon (1958). This was during the phase when France granted autonomy to its African colonies, but not independence. De Gaulle created a commonwealth of states called the Union Française (1956–60), which allowed for local assemblies to make laws about matters concerning the colony, but not giving them sovereignty (Think of overseas territories such as French Polynesia today.) De Gaulle had hoped to preserve his African empire, so important in his launching of the Liberation of France. In addition, Africa provided France with a continuing basis for its major power status in the international system during the Cold War.

De Gaulle's key adviser on African affairs was Jacques Foccart (1913–97), a white from the French Antilles who had joined the Resistance, and who ran an import–export business in the Empire after the war. In the 1950s he was an important financial contributor to the RPR, the Gaullist political party, and organized the SAC, an informal secret intelligence network of spies and commandos to defend French sovereignty and interests throughout the empire. He was involved in efforts to bring De Gaulle back to power in the murky events of May 1958, after which he became De Gaulle's shadow adviser on African affairs. (Péan 1983) It was he who pushed De Gaulle to break up the two large colonial federations of French West and Equatorial Africa into smaller countries before granting them self-rule. Foccart argued that it would be easier for France to control and dominate a dozen smaller African states than two large federations, especially in competition with the United States or other nations. He thought that Gabon, where petroleum resources were being developed at this time, would be easy to control because it was small, and because it lacked roads.

In 1960 he was named General Secretary of African and Malagasy Affairs, with an office in the presidential palace created to conduct French relations with newly independent African states. He simultaneously carried on clandestine operations through his "network," known as the *réseau Foccart*. This clandestine organization comprised old comrades from the Resistance, as well as hardened soldiers and mercenaries coming out of the bloody colonial wars in Indochina and North Africa. Its operatives often worked under cover as employees in his import–export business. Others had official posts in technical cooperation. With agents at every port watching and listening as his eyes and ears, Foccart masterminded numerous coups against independent-minded nationalists while installing in power a new generation of African rulers loyal to De Gaulle and himself. In the case of Gabon, he orchestrated the victory of Léon Mba as the country's first president. If Mba tried, at first, to promote real autonomy for his country, after a failed coup in 1964, when French paratroopers restored him to power, the president of Gabon thereafter took his orders from Foccart. This was how neocolonialism was consolidated in Gabon. (Péan 1983)

Here is the context in which Omar Bongo came to power. Gabon was an African country, geographically shaped by the French, colonized by the French, forested by the French, led by a francophone assimilated elite who spoke French, read French writing, received French education, practiced French law, worked for French businesses, and who had adopted a French system of government that was additionally dominated by France through a system called "'cooperation" (a term coined by the French in a series of military, economic, and diplomatic accords). Bongo's rise to power, and his record 42 years in office, were the result of his life-long collaboration with the French.

Omar (*né* Albert-Bernard) Bongo was born in Lewai (now Bongoville). a tiny village then located in the French colony of Middle Congo. The youngest of nine children, poor, unable to attend school until his father's death, he was sent to live with his oldest brother in Brazzaville. After finishing school he joined the French colonial army, and was posted to Chad, the Central African Republic, and the Congo. In 1960 blacks who had served in the colonial army were sent back to their countries of origin. Bongo found himself in a Gabon he did not really know. Fortunately he had made friends in the Foccart network who got him a posting in the new Foreign Ministry of Gabon, and he was soon promoted to the staff of President Mba. When the president became ill with cancer and flew to Paris for treatment, Jacques Foccart paid frequent visits to his deathbed to pressure the ailing old man to change his country's Constitution to introduce a post of vice-president, and to appoint Bongo as his running mate. Bongo was a complete unknown to Gabonese voters in the elections of 1967, so Foccart paid for radio and press slots, flying the young Bongo around the country (still lacking basic roads) to give speeches legitimating Mba's running mate. When President Mba died in 1967, Vice-President Bongo was installed in power. (Gardinier & Yates 2006)

Bongo remained loyal to the men who brought him to power. He admired Charles de Gaulle, and openly declared his gratitude to the French for their assistance in the development of his country. He cultivated strong personal ties with all subsequent presidents of France and worked closely with Foccart and his network, who exploited the vast natural resources of Gabon for the enrichment of France. The influence of France cannot be overemphasized. French remains not only the official language of Gabon, but also the spoken language of the majority of its people, who read French newspapers, listen to French radio, watch French television, study French literature and history in their schools, send their children to France for higher education, and have adopted French habits and tastes, including the importation of almost all of their food from France. In addition to its cultural imperialism, France has been the principal aid donor, lender, and trading partner. French businesses have dominated the industrial, commercial, and financial sectors. Through its "cooperation" accords France runs the Gabonese currency (the CFA franc), and thus makes Gabonese monetary policy. But the domination by the French companies of Gabon's oil industry is the focus of this chapter, so it is to that sector that we shall now turn our attention.

FRENCH DOMINATION BY ELF-GABON

The existence of petroleum in Gabon, both onshore and offshore, had long been known to the French. In 1929 geophysical crews from the Compagnie Générale Géophysique (today Schlumberger) had demonstrated its presence with electrical prospecting. But the French did not invest capital in developing these discoveries, in part because their privately owned national oil champion, Compagnie Française des Pétroles (today Total) was purchasing abundant cheap oil from its partners in the Middle East, and in part because the French state lacked the capital and technology in the 1930s to exploit colonial oil by itself.

Only after the Second World War did the French state develop that capacity and capital, under the aegis of the Bureau de Recherche du Pétrole (BRP), which financed a colonial oil company, Société des Pétroles d'Afrique Equatoriale Française (SPAEF), to conduct drilling operations in Gabon and the Middle Congo. SPAEF purchased a heavy rig from the United States and shipped it across the Atlantic to Port-Gentil on Mandji Island. The first wildcat wells were drilled in 1947 around Port-Gentil, and by 1951 a small oil reservoir had been found. Two more rigs were brought in to determine whether the wildcat was an isolated find or a trend. For the next five years SPAEF mapped out a trend of one hundred shallow salt-dome structures that ran up the spine of Mandji Island. While individually each inverted dome contained relatively small amounts of oil, collectively they represented a viable reserve. By 1957 eight SPAEF rigs were pumping enough crude to justify construction of a nine-mile-long oil pipeline to Cap Lopez, where the first cargo of "Mandji" crude was loaded for shipment to Le Havre. This was the beginning of the Gabonese oil industry. (Yates 1996: 56–7)

To expand oil production, new investments were needed. The whites in Gabon had little capital to invest, and the capital of the French state was heavily committed in North Africa. France therefore changed its policy of reserving overseas oilfields for French investors and sought foreign partners to promote more rapid development. Pierre Guillaumat (1909–91), French oil tsar, the boss of the BRP and the founder of the new French state oil company, Société Nationale Elf-Aquitaine, enjoyed good personal relations going back to his days in the Resistance with certain directors of Shell. So he invited that British oil firm to create a local subsidiary, Shell-Gabon, to work in the southern part of the country in partnership with the subsidiary he controlled, Elf-Gabon. Arab nationalization of Saharan oil pushed Guillaumat into shifting investments from the oilfields of the Sahara down to the Gulf of Guinea, where Gabon became the base of operations from which Elf constructed a veritable French oil empire that spread into the shallow waters offshore before moving southward to the Middle Congo, then northward to French Cameroon. (Yates 2009: 204–7)

During the late 1960s petroleum became Gabon's leading export and source of government revenue. By 1967 spectacular growth had occurred with the coming into production of the Gamba-Ivenga onshore south of the Ndogo Lagoon and the Setté-Cama, and of Anguille offshore just south of Port-Gentil. Port-Gentil became the site of the refinery of the Société Gabonaise de Raffinage (SOGARA), which was Gabonese in name only. A second refinery was opened nearby at Point-Clairette. Over the decade production increased eightfold from 1.4 million tons in 1966 to 11.3 million tons in 1976, with offshore accounting for 80 percent of this total. Eighty-five percent of the production was exported, the remainder being used in Gabon and its francophone neighbours. (Gardinier and Yates 2006: 261)

The transformation of the Gabonese economy from a forest enclave to an oil enclave resulted in greater Gabonese interest in petroleum, although true "Gabonization" of the sector was never achieved. During the 1970s the state did acquire a portion of the shares in the leading oil companies and a larger percentage of the revenues from production. In the expectation that these reserves would be exhausted by the end of the 1980s, the Bongo government gave the companies incentives to undertake new prospecting for additional deposits. But a downturn in world demand for oil began in 1977, resulting in a decade of financial crisis for Gabon. A spike in world oil prices in 1986 briefly filled state coffers with windfall profits, and oil rents generated five-sixths of export earnings and two-thirds of the government budget. Then a disastrous decline in oil demand, coupled with a weak dollar, set in motion a second crisis, deeper and more prolonged.

Oil was concentrated in four areas. The first was Port-Gentil, the oil capital, controlled by Elf-Gabon, and accounting for 76 percent of the national production. The second was Gamba, controlled by Shell-Gabon (also partly state owned), accounting for 8 percent of national production. The third was Maymba, also owned by Shell-Gabon, accounting for 7 percent. The fourth was Oguendjo, controlled by Amoco-Gabon, then owned by Standard Oil, providing the remaining 9 percent. In other words, after a quarter-century of political independence, the oil industry of Gabon was still dominated by the French company Elf, which controlled three-quarters of national production. (Gardinier and Yates 2006: 262)

A turning point in French domination came with the publication of *Affaires Africaines* (1983) by a French investigative journalist who revealed in great detail the workings of the Foccart network and the corruption of a business and politics it had cultivated. The revelations in this book caused a diplomatic crisis between France and Gabon and a scandalized Omar Bongo asked François Mitterrand to ban publication of the book in France, without success. From that point onward he started to look beyond France for new partners in development. Around this same time a major oilfield was discovered onshore by Shell Gabon. Rabi-Kounga field was of major importance to Gabon, because it provided half of Gabon's total output by 1991, and because Shell replaced Elf as the leading oil producer in the country. Of course, Elf ended up buying half of the equity, but Shell remained the operator. This huge discovery brought in European, American, Japanese, and even Korean investments to explore the area onshore and offshore. It also resulted in a change in the direction of oil flows, so that the largest share of oil exports were thereafter sold to the US.

Gabonese oil production peaked in 1997 at 364,000 barrels a day, and, lacking any new discoveries of the magnitude of Rabi-Kounga, it thereafter entered into a long, slow, and ineluctable period of decline: from 337,000 b/d (1998), 340,000 (1999), 327,000 (2000), 301,000 (2001), 295,000 (2002), 240,000 (2003), 235,000 (2004), 234,000 (2005), 235,000 (2006), to 230,000 (2007). (BP) Oil prices were extremely low during most of this period, around

20 dollars per barrel. The traditional actors, that is, the French and Anglo-American oil companies, were hesitant to invest in Gabon, where geological factors tended to make oilfields small and unprofitable. So they invested instead in horizontal drilling and sub-surface injection to extract more oil from the existing fields, thus allowing the petroleum sector to decline gradually toward ultimate exhaustion.

It was during this phase that another scandal broke out in the press, called the "Elf Affaire." While it was a public enterprise, Elf was the victim of considerable diversion of funds, carried out by the company's director of exploration and production, André Tarallo, who later confessed when the matter was brought to trial that retro-commissions from oil had been diverted to secret bank accounts in Liechtenstein, Luxembourg, and Switzerland, and to several dozen offshore accounts around the world. A few pennies on every barrel were put into these *caisses noires* systematically, and over several decades. Part of this embezzled money had been distributed to political actors in France, as well as to Omar Bongo and his collaborators, causing an enormous political scandal in Europe. The worst abuses were committed during the presidency of Loik Le Floch-Prigent (1989–93), when several billion francs were drained from the company treasury. This resulted in a financial weakening of Elf that would result in its acquisition by Total in 1999. But it also broke the neocolonial system by which Elf had dominated the oil industry in Gabon.

The aftermath of this political and economic scandal was the context in which the Chinese entered Gabon. Diplomatic relations between the two countries had been established in 1974, but it took 30 years for those relations to translate into tangible economic results. The truly emblematic event was the first official visit of a Chinese head of state to Libreville in 2004. During that historic trip, Hu Jintao and Omar Bongo signed an oil exploration and production agreement, and the chairman of China Petroleum and Chemical Corporation (Sinopec) led a business delegation to negotiate a "Memorandum of Understanding" for an oil sales contract between Total Gabon (formerly Elf) and subsidiaries of Sinopec and China International United Petroleum and Chemicals Company (Unipec) involving a million tonnes of crude oil per year. China has become the third largest importer of Gabonese oil, after the United States and France. (Yates 2008) President Hu also gave Bongo a cash grant of two million dollars and an interest-free loan of six million dollars, with "no strings attached." In addition to the crude oil purchases, Sinopec signed a technical agreement with the oil minister, Richard Onouviet, for three onshore oil fields around Port-Gentil; he explained that the agreement was designed to facilitate the investments Sinopec would have to make under a production-sharing contract because "prospecting for oil on land in Gabon is very difficult and costly because we are a heavily forested country." He implied that environmental regulations, which protected these forests as nature conservancies, would be relaxed: "We hope that Sinopec will discover a large deposit that is lying dormant under Gabonese soil." (Yates 2008: 209)

Despite the longstanding domination of France, despite its military, cultural, political, and economic influence, France is slowly losing control, and the Chinese have penetrated Gabon. The question that is now on everybody's mind is whether China will replace France there. The answer depends on what kinds of relations one believes really matter. If one examines only trade statistics, then it is entirely plausible that Gabonese exports could be sucked into the vortex of Chinese consumption. But if one considers how few Gabonese will ever learn Chinese, go to China, study there, write books in Mandarin, adopt Chinese lifestyles, settle there, and so on, then the crucial role of French cultural imperialism should become clear. Patterns that have taken centuries to form will not disappear overnight. In the larger scheme of things, Gabon is located in the Gulf of Guinea, geographically closer to Europe and America than to China. It plays a small but important role in Western security for central Africa, as the site of a French military base, and possibly a future American one. Only the conscious abandonment of the country by the French or the Americans would open the door to greater Chinese strategic influence. That being unlikely, in the present configuration, one should expect to find Chinese activities developing under the aegis of Western domination of Gabon, or, at best, under depoliticized globalization. But there is no reason to predict that China will establish a relationship anything like the neo-colonial system created by France, at least not in Gabon. For the present, Total is the dominant firm.

In 2009 Omar Bongo finally died, in a private clinic in Barcelona. For a week the government actually denied his death, worried that the news would provoke riots in the streets. Having stayed in office until his last dying breath, President Bongo had left a serious problem of succession for those he left behind. France wanted stability and, after a brief period of uncertainty, supported his son Ali Ben Bongo, who assumed the role of legitimate heir. Ali rallied different factions of the Bongo clan within the family dynasty, and won the nomination of the ruling PDG party. He quickly announced that elections would be held in August, in order to prevent the numerous opposition parties from having the time to settle on a common candidate. He refused to resign from his position as defense minister and declared a state of emergency, which effectively prevented the opposition from holding anti-Bongo rallies. His control of the state media, combined with the repression of all opposition media, ensured that his campaign would not be contested in the audiovisual world of the national consciousness. The elections were fraudulent, with over 2,000 claims of irregularities in different electoral bureaux. In the end he took the elections with a little over 41 percent of the vote, defeating former Interior Minister André Mba Obame and long-time opposition leader Pierre Mamboundou. Riots broke out in the streets of Port-Gentil, the oil capital, but were brutally repressed by the army. In the end his inauguration on October 16, 2009 indicated that stability would prevail over change, and Gabon would continue with business as usual.

2
Multinational Corporations and Nationalization

When you consider an African oil enclave, one of its single most striking features is the domination by and dependence upon foreign multinational corporations (MNCs) that own it. They hire their own exploration teams. They build their own offshore drilling platforms. They run their own pumping stations, pipelines, refineries, heliports and tanker fleets as they please. Their global distribution networks, world-class investments, and superior technology give them a kind of sovereign power over poor rural African villages located around the enclaves. For 50 years, many oil-rich mangrove swamps of the Niger Delta were governed by Shell Oil. Deepwater oil rigs in Gabon and Congo were governed by Elf as though they were free island archipelagos. In the Cabinda enclave of Angola, Gulf Oil decided what happened behind its barbed-wire fences. Today, Asian firms imitate the Euro-American "landlords" who rule over African concessions. PetroChina, Malaysian Petronas, and the Indian Oil and Natural Gas Company have each purchased concessionary rights, and now control vast territories in the Sahara and the Sahel. Despite what you may think, this is not the case in other regions of the developing world, such as Latin America, the Middle East, and East Asia, where leaders have successfully nationalized their oil sectors. So the question remains: Why has nationalization *not* succeeded in Africa? Why do *foreigners* still rule African oil?

THE LEGACY OF GEOGRAPHY?

In his Pulitzer-prize-winning *Guns, Germs, and Steel (1999)*, anthropologist Jared Diamond argues that different levels of technological development allowed Europe to conquer Africa (and, by implication, its oil), but Europe's comparative advantage over Africa ultimately derived from long-term consequences of differences in their environments. The different trajectories of Europe and Africa are "due to accidents of geography and biogeography—in particular to the continents' different areas, axes, and suites of wild plant and animal species." (p. 400) Africa is smaller than Eurasia and had fewer indigenous crops and animal species. Also, those species domesticated in one part had great difficulty moving to other parts of the continent because, unlike Eurasia, the major axis of Africa is north/south: the Sahara and Kalahari, the Congo, the Equator, and two tropics posed geographical barriers to crop migration from the north to the south. For the same reason, human technologies such as guns and steel arrived later in Africa than in Europe. "Guns and steel" were of course the proximate causes of European conquest. But the remote causes, Diamond concludes, were geography and environment.

Diamond's brand of biogeography is compelling, but his remote prehistoric factors are not sufficient to explain why other similarly situated oil-producing regions did in fact develop national champions. For example, South America had similar environmental

---

Box 2.1   Biogeography

"[Q]uestions about inequality in the modern world can be reformulated as follows. Why did wealth and power become distributed as they now are, rather than in some other way? For instance, why weren't Native Americans, Africans, and Aboriginal Australians the ones who decimated, subjugated, or exterminated Europeans and Asians?" (p. 15)

"Authors are regularly asked by journalists to summarize a long book in one sentence. For this book, here is such a sentence: 'History followed different courses for different peoples because of differences among peoples' environments, not because of biological differences among peoples themselves.'" (p. 25)

"Naturally, the notion that environmental geography and biogeography influenced societal development is an old idea. Nowadays, though, the view is not held in esteem by historians; it is considered wrong or simplistic, or it is caricatured as environmental determinism and dismissed, or else, the whole subject of trying to understand worldwide differences is shelved as too difficult. Yet geography obviously has some effect on history; the open question concerns how much effect, and whether geography can account for history's broad pattern." (pp. 25–6)

Jared Diamond, *Guns, Germs, and Steel: The Fates of Human Societies* (1997)

disadvantages of area, axis, and diffusion of domesticated plant and animal species, but today the Brazilian state oil company Petrobras dominates Brazilian oil. In 1953, Brazil's president decreed a state monopoly over all exploration and refinement, and founded Petrobras to exercise this monopoly. Petrobras became one of the world's hundred-largest corporations. (Evans 1979: 217) Pemex controls Mexican oil. Petronas controls Malaysian oil. PetroChina controls Chinese oil. Try to think of an African equivalent. Prehistory suffices to explain why Africans *were* dominated; but is insufficient to explain why their oil enclaves still *are*.

THE LEGACY OF SLAVERY?

Slavery is a second factor of African dependency. Here it is not possible to falsify by comparison with some other place that had a comparable historical experience. No region of the world was ever so intensely enslaved over such a long period of time as petroleum-rich West Africa. More than 7 million West Africans were exported in the Atlantic trade, constituting around 60 percent of that trade. If you include the Islamic trade, between 8 and 9 million slaves were exported from Africa over four centuries. (Drescher and Engerman 1998: 34) Estimates of the size of the slave trades remain a matter of contention among scholars. The most important long-distance trade was the infamous "Atlantic Triangle" between West Africa, the Americas, and Europe.

The number of people involved has been difficult to establish firmly. At least 11.7 million slaves were taken from Africa between AD 1450 and 1900. But around 15 percent to 20 percent died in the Middle Passage. Many others died on forced marches from the interior of Africa to the slave ships on the coast. There were also differences of trade volume over time. The slave trade rose and fell in size: Some 3 percent of total slaves were taken between 1450 and 1600, 16 percent in the 1600s, 52 percent in the 1700s, and 29 percent in the 1800s. (Higman 1998: 171) A second major slave trade was directed to the Muslim areas of North Africa, Arabia, and India. Arab slavers took African slaves across the Sahara, the Red Sea, and the Indian Ocean, with levels of mortality that probably rivaled those in the Middle Passage. But 1.9 million Africans crossed the Atlantic in the seventeenth century (65 percent of total), and 6.1 million in the eighteenth century (82.5 percent of the total). This Atlantic-oriented slavery, therefore, fundamentally shifted the direction of African trade from a Mediterranean-oriented trans-Saharan route to an Atlantic coastal commerce. Guns, rum, and gold coins in the European enclaves, traded for slaves from the interior, weakened the economies of inland African empires, and allowed European sea powers to establish empires premised upon pre-existing coastal bases. Slavery provided money for ports and forts, and prepared the way for Europe's "Scramble for Africa" (1876–1914).

Was the slave trade profitable at the firm level? What impact did it have on European economic growth? What impact did it have on African growth? The answer to the first question is that profits in the slave trade were not extraordinary by European standards. The average 10 percent profit rate obtained was considered a good rate at the time. To the second question, although key industries may have relied heavily on the African trade, it is now the consensus that British industrial growth was not financed by profits from the slave trade. To the third question, the slave trade exacerbated African warfare. It promoted the spread of epidemics across the continent. It depopulated large zones of potential agricultural growth. It extraverted African coastal trading societies and explicitly reinforced their export-orientation. In a sense, it prepared the way for the African oil trade.

CAPITALISM? NEOCOLONIALISM? AUTHORITARIANISM? COLLABORATION?

A third factor of African dependency is imperialism. Africa's weak structural positioning in the international capitalist system explains its dependency on foreign firms. Dependency-and-under-development theorists such as Dos Santos (1970), Amin (1976), and Frank (1979), believed that the cause of African poverty was the capitalist "world-system." (Wallerstein 1974) They saw the modern world system as divided into a rich capitalist core and poor underdeveloped periphery (imagine two concentric circles). Five centuries of domination, through the ages of discovery, exploration, conquest, slavery, colonialism, and imperialism, were structured by unequal rates of *capitalist development*. The capitalist core always possessed military, economic, and technological superiority. It reinforced this inequality through unequal trade of cheap raw materials for expensive value-added goods. As this trade increased, so did inequality. Capital accumulated in Europe. Slaves accumulated in the Americas. Nothing accumulated in Africa. In Walter Rodney's

How Europe Underdeveloped Africa (1974), "underdevelopment" is used as a verb, and not a noun.

Dependency theorists blamed the "world system" for African poverty, but did not explain how wealth came to Persian Gulf Emirates or the East Asian Tigers. The world system is a common factor to all countries in the periphery, so it is not sufficient to explain different outcomes in different regions of the periphery. Cardoso (1978) and Evans (1979) reported two major economic success stories in the Latin American periphery (Brazil and Mexico), and challenged the assumption that dependency always causes under-development. On the contrary, they argue, in Brazil and Mexico, capital *has* in fact accumulated, with development stimulated in the process. States such as these, which industrialized under conditions of dependency, illustrate a new kind of growth that Evans called "dependent development." But if Latin American oil industries are owned and run by large, integrated, national oil companies— Petrobras (Brazil), Pemex (Mexico), and Petróleos de Venezuela (Venezuela)—why is there no African equivalent?

The crucial stage was not the prehistoric era, as Diamond suggests, but the *post*-colonial. African oil companies are more dependent because political independence came later; and when it came, it had evolved into "neo-colonialism," a fourth factor of African dependency. Foreign powers took advantage of the weak

---

Box 2.2   Capitalism and colonialism

"Under colonialism, communal ownership of land was finally abolished and ownership of land imposed by law. [...] With the seizure of the land, with all its natural resources and the means of production, two sectors of the economy emerged—the European and the African, the former exploiting the latter. Subsistence agriculture was gradually destroyed and Africans were compelled to sell their labor power to the colonialists, who turned their profits into capital. With the growth of commodity production, mainly for export, single crop economies developed completely dependent on foreign capital. The colony became a sphere for investment and exploitation. Capitalism developed with colonialism." (pp. 14–15)

Kwame Nkrumah, *Class Struggle in Africa* (1970)

institutions of newly independent post-colonial states. African oil may have been discovered during the colonial era, but it was mostly exploited *after* decolonization. Britain may have discovered Nigerian oil in the 1920s, for example, but it produced it mostly after independence. France discovered Gabonese oil in the 1950s, but produced it mostly after independence. The same is true for US oil production in Angola, and Asian oil production in Chad.

Neocolonialism explains postcolonial dependency as the effect of continued economic dominance by foreign powers. Yet decolonization was not uniform, nor was neocolonialism a necessary outcome. While France did construct a vast cooperation network to hold on to power in its former colonies in Africa, France was the *exception*, and not the rule. In the Commonwealth, for example, Britain did not create a "Pound zone," or a specifically targeted "Africa policy" like the French. (Britain instead has a "Commonwealth policy" including African member states.) *A fortiori*, Portuguese, German, Italian, and Spanish colonizers in Africa never became neocolonial powers as did the French. Portugal's defeat in its long anti-colonial wars left it with little more than a linguistic sphere of influence, while Germany, Italy, and Spain left Africa with barely a tangible interest at all. Even the French sphere of influence—"*Françafrique*"—is now in decline. This view is widely held by French Africanists, such as Antoine Glaser and Stephen Smith, who proclaimed in their best-selling paperback *How France Lost Africa* (2005) that the old days of French influence are gone: "*Françafrique est morte.*" What remains is African extraversion.

"*Extraversion*" is a term coined by the French political scientist Jean-François Bayart (1993) to describe how relatively weak economic development, and internal social struggles, made African political actors more "extraverted," that is, more disposed to mobilize resources from relations with actors in the international system; our fifth factor of African dependency. African extraversion was conditioned by a traditional political culture that Bayart called the "politics of the belly" (a cultural attitude towards power and consumption expressed by the African folk saying, "A goat grazes where it is tethered"—that is, a ruler will consume wealth where he is sovereign). One indicator of extraversion is the ratio of trade to national income. Trade extraversion is greater in Africa than in any other region of the world. Glaser and Smith (2005: 24) remarked, "The proportion of African trade destined for other regions of the world is 45 percent of GDP"—more than Europe (12.8 percent), North America (13.2 percent), Asia (15.2 percent), or even Latin

Table 2.1 Rulers of African oil-rentier states (2009)

State (N = 10)	Ruler	Years in Power	Profession	Ruling Parties (Seats/Total)
			"Democrats"	
Nigeria	Goodluck JONATHAN	(0) 9 Feb 2010	University professor, elected Vice President, now Acting President.	72% (260/360)
São Tomé	Fradique De MENEZES	(9) 3 Sep 2001	Businessman, former finance minister. Elected with minority in parliament.	42% (23/55)
			"Françafricans"	
Gabon	Ali BONGO	(1) 16 Oct 2009	Politician, former defense minister. Son of the previous president.	72% (86/120)
Cameroon	Paul BIYA	(28) 6 Nov 1982	Political scientist installed by French. Successor to the founding father	78% (140/180)
			"Praetorians"	
Angola	Edouardo DOS SANTOS	(31) 21 Sep 1979	Professional solider. Former communist dictator 1979–91. Won a civil war.	87% (191/220)
Equatorial Guinea	Obiang NGUEMA	(31) 3 Aug 1979	Professional solider. Came to power by coup. Killed his uncle the founding father	89% (89/100)
Congo	Denis SASSOU NGUESSO	(26) 25 Oct 1997	Professional solider, dictator 1979–92. Returned to power in 1997 civil war	85% (56/66)
Sudan	Omar AL-BASHIR	(21) 0 Jun 1989	Professional solider. Came to power by coup. Consolidated power in a civil war	79% (355/450)
Chad	Idriss DEBY	(20) 4 Dec 1990	Professional solider. Came to power by coup. Survived several attempted coups	71% (110/155)
Mauritania	Mohamed AZIZ	(2) 6 Aug 2008	Professional solider. Came to power by coup. Deposed democratically elected president.	N/A
Average		13.8 yrs	60 % Professional Soldiers	75% (N = 9)

America (23.7 percent). Extraversion, they argue, keeps the postcolonial state in Africa dependent on former colonizers, on rising commercial powers, on major foreign enterprises, and on international financial institutions such as the World Bank or the IMF. When looking at worker remittances, tourism, communication by telephone or internet, import/export of books, films, and other cultural commodities, the globalization of Africa is even greater.

A sixth explanation is that "African oil regimes are corrupt dictatorships." Nine out of ten rulers in African oil-rent dependent states call themselves "President," but only two came to power through democratic elections: Goodluck Jonathan and Fradique De Menezes. Their "Françafrican" counterparts, Ali Bongo and Paul Biya, came to power by means of fraudulent plebiscites orchestrated by the French. The remaining six military "praetorian" rulers—Obiang, Dos Santos, Sassou-Nguesso, Al-Bashir, Déby, and Aziz—are professional soldiers who seized power by coup d'état and/or war and keep it through the selective use of violence (see Table 2.1). Looking at the grim collection of patriarchs, paratroopers, and fighter pilots who preside over 80 percent of our cases, are we seeing the veritable problem (authoritarianism), and its solution (democracy)?

No. While democracy may be compatible with capitalism, the idea that certain freedoms, notably economic freedoms, are beneficial to wealth creation goes back to Adam Smith (1723–90). And while democracy protects property rights, and it is easier to trade with other democracies if one conforms to international democratic standards, *democracy does not prevent dependency*. Nigeria is a democracy, yet foreign oil corporations still dominate its oil industry (see Chapter10). São Tomé is also a democracy, and it is equally dependent on foreign multinationals (see Chapter 7). Nothing about democracy *per se* makes it impossible for foreigners to continue to dominate Nigerian or São Toméan oil, given surplus capital, and a thirst for direct foreign investment. Both dictators and democrats collaborate with foreign oil companies.

This brings us to our last factor: the corruption of African rulers. For whether or not a ruler is democratic or authoritarian, what matters is whether he collaborates or is corrupt. Robinson's (1972) "theory of collaboration" suggests that foreign domination derives as much from indigenous collaboration, the corrupting influences of money and power, as from the behavior of exogenous actors. Collaboration is a necessary cause of African dependency. Geography, slavery, capitalism, neocolonialism, and authoritarianism may be sufficient, when combined, to explain African dependency, but none of them is necessary to that argument. Remove any one of them, and you still have the three others to replace it, either by themselves (e.g. slavery is sufficient), or combined (e.g. slavery and neocolonialism). But collaboration is a necessary cause, the *sine qua non*, without which whites would not have acquired their oil concessions in Africa. Remove collaboration, and you remove the effect.

Box 2.3   Theory of collaboration

"[Imperialism's] controlling mechanism was made up of relationships between the agents of external expansion and their internal "collaborators" in non-European political economies. Without the voluntary or enforced cooperation of their governing elites, economic resources could not be transferred, strategic interests protected or xenophobic reaction and traditional resistance to change contained. Nor without indigenous collaboration, when the time came for it, could Europeans have conquered and ruled their non-European empires. From the outset that rule was continuously resisted; just as continuously native mediation was needed to avert resistance or hold it down." (p. 120)

"The theory of collaboration suggests that at every stage from external imperialism to decolonisation, the working of imperialism was determined by the indigenous collaborative systems connecting its European and Afro-Asian components. It was as much and often more a function of Afro-Asian politics than of European politics and economics." (p. 138)

Ronald Robinson, "Non-European Foundations of European Imperialism," (1972)

In the end, European imperialists lost their old monopolies in their oil spheres of interests to newer, American, and now Asian commercial empires. Africans have found new partners, and will continue to find them, because the world is thirsty for African resources, but also because, as Robinson's "theory of collaboration" suggests, at every stage, the working of imperialism was determined by the indigenous collaborative systems connecting its European and African components: "Imperialism was as much a function of its victim's collaboration or non-collaboration of their indigenous politics as it was of European expansion." (p. 140)

Map 2.1   Cabinda enclave, Angola

CASE STUDY: YANKEE LANDLORDS OF CABINDA

Why has the Cabinda Enclave of Angola been owned and run by American corporations? Of course, the United States is the world's most powerful country, and Angola is sub-Saharan Africa's largest oil producer. With estimated reserves of seven to ten billion barrels of oil, Angola is located less than half the distance of the Persian Gulf from American consumers. So geography and American national interests can explain American involvement in Angola. Also, over 97 percent of Angola's oil is drilled offshore. (Gary and Karl 2003: 31; Hodges: 126) Since only large integrated corporations, with substantial capital and technology, are capable of drilling major deepwater offshore reserves, foreign multinationals have always dominated. But why *American* multinationals? It is important to understand that, unlike aid-dependent African regimes, which find it hard to resist their benefactors, Angola has been blessed with abundant natural resources that free it from strings of conditionality attached to foreign aid. Its military struggle for independence (1961–75), and its long civil war (1975–2002), have also made the MPLA resistant to foreign powers—like the US—which once armed its enemies. Angola is free to choose its partners. Moreover, specialists have increasingly argued that the national oil company Sonangol now has the technical and organizational capacity to regulate the industry and become involved in oil production, distribution, support services, businesses, banks, and refinement (Soares de Oliveira 2007). So why is the oil-rich Cabinda Enclave *not* owned and run by Sonangol?

Portugal was the first European power to reach Angola, over five hundred years ago. The earliest Portuguese contact with the peoples of Angola occurred in 1483, with the discovery of the native kingdom of Kongo at the mouth of the Congo River. At that historical time period, Portugal was one of the most advanced sea powers in the world. As a consequence, the king of the Kongo, who traded ivory and slaves for guns and other modern products, accepted a Portuguese adviser in his court, converted to Catholicism, invited Christian missionaries to his country, renamed his capital São Salvador, and even changed his name to the Portuguese "King Affonso." (These are all examples of African extraversion.) King Affonso's initial collaboration with Portugal resulted in Angola becoming an important source of slaves. Portugal founded the settlement of Luanda in 1576 to facilitate the trade. It has been estimated that between 1580 and 1836 Portugal enslaved over 3 million Angolans. (Minter 1972: 22) Most crossed the Atlantic, establishing an Atlantic-oriented pattern for trade.

This pattern did not result in control by Portugal itself, however, because of the political geography of the Atlantic trade. By the sixteenth century, the Portuguese empire was economically centered on Brazil and Angola: two Atlantic colonies tied together by a booming slave trade. Portuguese colonists in Angola furnished workers for the sugar plantations of Portuguese colonists in Brazil, whose export revenues provided Portuguese colonists in Angola with the money they needed to finance their colony. Angolan slaves were also transported to the United States, beginning a long history of trade between those two nations. After slavery was abolished in 1815, tens of thousands of Angolan slaves were still being illegally shipped to Atlantic islands in the Gulf of Guinea and Caribbean. But looking at the total trade figures, slavery established an American-oriented flow of Angolan exports. This can be explained by the fact that, despite its longevity, the intensity of Portuguese colonization in Angolan territory only increased after the Scramble for Africa (1876–1914). Its historical demise during the Napoleonic Wars had reduced Portugal to a poor, weak, British client state. In fact it was not until the twentieth century that Portugal's colonial conquest of the Angolan interior could be finished, and significant numbers of Portuguese colonists arrived.

In 1950 the white settler population numbered only 78,826 (Somerville 1986: 20), and the colonial economy, based on the slave trade and then on forced labor, depended on native labor, which had low overall levels of human development. After independence came in 1975, there were 340,000 white settlers in Angola, mostly new arrivals, who managed the businesses and ran the colonial bureaucracy. In 1975 these white settlers suddenly left, in a matter of days, causing the newly independent People's Republic of Angola to suffer a lack of what is called "human capital" (individual knowledge, skills, and capabilities). Low levels of human capital contributed to Angola looking elsewhere for industrial partners.

No discussion of collaboration in Angola would be complete without some discussion of Portuguese racism, which created categories of culturally assimilated blacks (*assimilados*) and mixed-blood creoles (*mestiços*), new peoples who were generated by five centuries of commerce, missionary education, and inter-racial marriage. These Portuguese Africans occupied a middle status between the white settlers and the indigenous blacks (*indigenas*), who were at the bottom of the social order. One of the main characteristics of Portuguese colonial rule was the racial mystique created about its colonies. Central to this mystique was the idea of a "pan-Lusitanian" community unified by Portuguese language and culture. The theoretical basis for colonial policy became known as "Lusotropicalism," which emphasized Portugal's unique ability to blend with indigenous peoples around the world, and to assimilate them into Portuguese civilization. Assimilated Africans in 1950 numbered around 30,000 Angolans. Six years later a colonial report showed 68,759 Africans attending school, being assimilated. This select group of Portuguese-speaking Angolans received special treatment. Without them, the new white European settlers from Portugal could not have implanted their plantations, nor put down uprisings. For the *assimilados* and *mestiços* managed indigenous African labor. They worked as overseers for white businessmen, collaborated with the colonial administration, served in the colonial militia, spoke the language of the colonizers, wore their clothes, and adopted their outlook. They were a necessary condition. Without them, Portugal could not have ruled Angola.

PORTUGUESE COLONIALISM, AMERICAN IMPERIALISM

Portugal's colonial contradiction was that, one the one hand, it remained economically dependent on the United States, but on the other, it sought to maintain its African empire. Portugal's fascist dictator, Antonio Salazar (1889–1970), was an infamous Cold War collaborator with the Americans. Originally an ally of Mussolini and Franco, he changed sides during the Second World War, and negotiated a treaty with the US to build an air based on the Azores islands (1943). George Kennan, then *chargé d'affaires* in the US embassy in Lisbon, wrote a letter to Salazar explaining, "The United States of America undertakes to respect Portuguese sovereignty in all Portuguese colonies." (Minter 1972: 39) The US allowed Portugal to join NATO as a founding member (1949), with US commitments to Portugal embodied in the new treaty. Military assistance from the United States to the Salazar regime increased from $0 in 1950, to $71 million in 1953. US military assistance meant that substantial portions of the Portuguese colonial armed forces were American-equipped and American-trained. This solved Portugal's contradiction. Portugal was weak, yet held Angola by force. "Africans in Portuguese Africa have increasingly recognized that they fight not only against Portugal," wrote William Minter, "but against a whole imperial complex headed by the United States." (1972: 99)

At the heart of the matter were American attitudes toward the Cold War and, especially, toward the Soviet Union. American policy on Angola had been formulated almost exclusively around the questions of what signal it would send to Moscow. John Foster Dulles, the principal architect of Eisenhower's foreign policy, considered Portuguese colonies part and parcel of Portugal, and perceived African nationalism as a tool of Moscow's creation rather than a natural outgrowth of the colonial experience. John F. Kennedy saw nationalism as the wave of the future in Africa, but he believed that the longer the wars for liberation endured, the greater the opportunity for the Soviets to gain influence over nationalist parties. The Johnson administration attempted to pursue a two-track policy of appealing to both Portuguese and Angolan nationalists, but in the end the strategic military interests in Europe were more important than those in Africa.

American involvement in Angola was a constant in the twentieth century. The US was a Cold War ally of Portugal during the Angolan war of independence (1961–75). Then the US was an ally of the FNLA and UNITA during the long civil war (1975–2002). After the end of the Cold War the US became the ally of the MPLA. At each stage in Angolan history, the US always found new collaborators for its investments, aid, and trade. It is to these that we must turn to understand why a US company controls Cabinda oil.

Before 1960, much of the Portuguese economy was foreign-controlled. But American direct investments in Angola had been comparatively minor, consisting mainly of oil exploration contracts signed by Gulf Oil. It was primary agricultural commodities that were the most important sector of the colony, accounting for some 60 percent of the value of total exports. Coffee was Angola's most important commodity, accounting for approximately 50 percent of export revenues alone. In those days, Angola was the world's fourth largest coffee exporter. Around half of that coffee was shipped to America, with the US accounting for around one-fourth of the colony's coffee exports. (Minter 1972: 123) But Americans had not *directly* invested in Portuguese coffee plantations. Major expansion of direct US investment in Angola only came *after* the beginning of Angola's struggle for liberation in 1961. In particular a new law opening Portugal's overseas possession to foreign direct investment resulted in major direct US investment in Angolan oil. In 1965, Salazar desperately needed revenues to fund three wars against anti-colonial guerrillas in Guinea, Mozambique and Angola, so he decreed a law that made possible, for the first time, foreign direct investment in Portuguese overseas possessions without any participation by Portuguese capital. (Minter 1972: 116) American multinationals took advantage of this new opportunity, and increased their equity in the oil concessions.

Gulf Oil had begun its exploration in Angola in 1957, but would not strike oil in Cabinda until 1966, after Salazar's new decree. Gulf became the sole concessionaire in the enclave. Before the discovery of oil in Cabinda, annual production of Angolan oil had been 650,000 tons. With the addition of Cabinda, oil production rose to 1.5 million tons, and was predicted to rise to 7.5 million tons by 1970, then perhaps to 15 million tons by 1973. (Minter: 119) The Portuguese colonial governor-general,

Rebocho Vaz, expressed the importance of Cabinda: "As you know, oil and its derivatives are strategic minerals indispensable to the development of any territory; they are the nerve centre of progress, and to possess them on an industrial scale is to ensure essential supplies and dispose of an important source of foreign exchange" (p. 120). Oil revenues paid to the Portuguese government were substantial, given the size of Angola's economy. Payments by Gulf Oil included surface rents, bonuses, income tax, royalties, and concession payments. According to the company, Gulf paid $11 million to the government in 1969. This equaled half of the colonial defense budget (ibid.).

Not only was Gulf Oil an important source of revenues for the colonial government, but oil was also a strategic natural resource for the infrastructure of modern war. "Machines cannot move without fuel," remarked Vaz. "Hence the valuable support of Angolan oil for our armed forces" (ibid.). Therefore Gulf's arrangement with the Portuguese government also included provisions for cooperation in defense of the zone in which the oil was produced. Portugal agreed "to take such measures as may be necessary to prevent third parties from interfering with the company's free exercise of its contractual rights." In return, Gulf was required to help provide for its own defense, with its activities fitted into the government's civil defense structure. As the enclave's oil potential grew, it became an important source of revenue, and so Portugal fought even harder to keep it.

ANGOLAN NATIONALISM AND THE MPLA

The key to African nationalism is the empowerment of leaders who do not collaborate. Although resistance to Portuguese rule had existed ever since the first attempts were made to establish a colony, it had usually taken the form of more or less spontaneous, defensive uprisings, or scattered revolts sparked off by particular instances of oppression. Those efforts at resistance were brutally repressed by the white settlers and their *assimilados* collaborators. Angolan nationalism, as a coherent set of ideas with an organized movement to promote it, did not really surface until the 1950s, when a small number of radical, educated *assimilados* (including Agostinho Neto) established a Center for African Studies in Lisbon. The aim of the center was to "evoke the sense of belonging to an oppressed world and awaken a national consciousness" through studying African cultures (Davidson 1975: 155). The Salazar regime suppressed the center, but another more clandestine movement was founded, some say in 1953, others in 1955. This coalition established a new, revolutionary, "nationalist" political party called the *Movimento Popular de Libertaçao de Angola* (MPLA). The Popular Movement for the Liberation of Angola was a nationalist front aimed at uniting efforts in the struggle against Portuguese colonialism. In the 1950s the colonial government waged a campaign of arrests and forced exile attempted to crush it. Dr. Agostinho Neto was himself arrested several times (in 1951, 1952, 1955, 1957, and 1960) and his MPLA headquarters were forced into exile, first in Paris, then in Conakry, then in Leopoldville.

It was the MPLA raid on the central prison in Luanda in 1961 that opened the War of Independence (1961–75). This raid coincided with the presence of a large number of foreign journalists, so it was impossible to conceal. The MPLA hoped to free a number of its political prisoners who were about to be executed, and although those prisoners were not freed, the violence served as a catalyst throughout the country. A rebellion broke out over wide areas of northern Angola, but the MPLA leadership was scattered in prison and exile, and it lacked a base from which to launch attacks. It was unprepared for a long guerrilla war. When the MPLA was expelled from ex-Belgian Congo (Kinshasa) by US collaborator Joseph Mobutu, it had to move its headquarters once more, to the ex-French Congo (Brazzaville). From there Neto naturally concentrated his military operations on the neighboring enclave of Cabinda. His guerrillas successfully occupied more than 90 percent of the enclave, thoroughly routing the Portuguese militia in the mountains and in the roadless tropical rainforest of Cabinda. The result was that MPLA had taken Cabinda from Portugal before Angolan independence, but stopped just short of taking the provincial capital, Cabinda City, where Gulf Oil had its main operating facility, Malongo. At the height of the raids, Gulf Oil was forced to suspend its operations, but it soon moved back in as Portugal re-established some control.

Portuguese efforts in Cabinda were particularly intensive after the discovery of oil in 1966. Cabinda received the largest allocation for colonial concentration camps, "rural regrouping projects," which the colonial government also euphemistically designated "strategic hamlets." In these camps the people of Cabinda became civilian victims, hostages of 14 years of hostilities between two "foreign" armies. When the Portuguese finally left Angola in 1975, MPLA soldiers occupied Cabinda immediately and declared it a part of

Angola. At this historic moment in Angolan history, the question is simply this: Why didn't MPLA take the offshore concessions away from Cabinda Gulf Oil Corporation? This is the crucial historical phase in Angolan oil nationalism, when Gulf's colonial concessions could have been revoked. The fall of Lisbon, the end of Portuguese colonialism, and the arrival of a newly independent People's Republic of Angola resulted in the "nationalization" of Angolan oil. So why did Gulf Oil still run Cabinda? Why does Chevron run it today?

The answer lies partly in the geopolitical realities of Cabinda itself. Unlike Angola, which had been a Portuguese colony since the fifteenth century, the Cabinda Enclave had been a triumvirate of kingdoms (Kakongo, Ngoyo, and Loango), largely autonomous until 1885, when they sought Portugal's protection against the Belgian and French Congo on either side of them. In 1885 these kingdoms signed the Simulamboco Treaty and became a Portuguese protectorate. Thereafter the Cabinda Enclave was known as the "Portuguese Congo." The Cabindans were Kongo people. They were Portuguese-speaking only by an accident of history, and were closer in culture, if not in language, to their French-speaking neighbors in Brazzaville and Leopoldville. Cabinda is home to barely two percent of Angola's population, and occupies less than one percent of the total surface area. Two hundred thousand people live in all of Cabinda, many of them in remote mountain villages. The small population and territory environmentally conditioned their domination by foreigners. Large oil reserves also made their enclave vital to Angolan national strategy, in the same way that it had been vital to Portuguese colonial strategy. From the perspective of the Cabindan people, however, the old white Portuguese settlers had been replaced by new Angolan nationalists, whose leaders were *assimilados*, *mestiços*, or came from other ethnic groups, notably the Mbundu, around Luanda, and whose "nationalization" was really a new form of internal colonialism (see Ukiwo 2008).

Cabinda's struggle against MPLA rule began well before the really large-scale oil production got under way off the coast. The struggle can be dated to 1963 with the founding of the Front for the Liberation of the Enclave of Cabinda (FLEC), a guerrilla independence movement fighting to free the enclave of both Portuguese and Angolan colonizers. FLEC guerrillas opposed any attachment of their enclave to Angola. Cabinda was independent. Thinking counterfactually, if Cabinda had been allowed to become independent, then its citizens would today be among the richest

African nations per capita. But thinking factually, as the province's oil potential grew it became an important source of revenue for Angola, which fought even harder to retain sovereign control. Oil revenues from Cabinda became the principal source of revenue for the MPLA during the long civil war (1975–2002), accounting for over 90 percent of their government's budget. Tax receipts from Cabinda to the Angolan National Treasury totaled $2.5 billion (2000) or 71 percent of government revenues. Some of the civil war's most brutal battles were waged between the MPLA and FLEC, until they signed a peace accord in July of 2006.

During that time the enclave was sold, and re-sold, to foreign investors by an occupying force of purportedly "Marxist-Leninist" soldiers. Malongo operating facility, on the outskirts of Cabinda City, was originally built by Gulf Oil, but was later acquired by Chevron. In 1984, during hard times in the world oil markets, Gulf Oil was bought by Chevron (now a part of ChevronTexaco). Malongo is today surrounded by an electric fence, razor wire, and several hundred landmines, visible reminders of 41 years of total war. The landmines were laid by Cuban soldiers under orders from the MPLA to protect Gulf Oil's operations against UNITA. John Ghazvinian, who visited Cabinda (and didn't get arrested) has reported that Chevron's foreign personnel are picked up in buses when they arrive at Cabinda airport and driven straight to Malongo, before they are flown by helicopter to the offshore oil platforms. While waiting at Malongo they live in an enclave-within-an-enclave, a "cocooned existence of satellite TV, basketball courts, imported American snacks, and even a rolling golf course." (Ghazvinian 2007: 154). This isolated enclave of luxury, surrounded by the larger enclave of misery, has produced a deep sense of grievance among the Cabindan people, who complain that Chevron discriminates against them and prefers to hire expatriates.

Technically speaking, Gulf Oil's concession in Cabinda (the famous "Block 0") is a legal construct of the *Petroleum Law N° 13/78*, passed by the regime on 26 August 1978. Article I of this historic oil-nationalization law declares:

[A]ll deposits of liquid and gaseous hydrocarbons which exist underground or on the continental shelf within the national territory […] or within any territory or domain over which Angola exercises sovereignty […] *belongs to the Angolan people*, in the form of *state property* (Article I, italics added).

What is important to understand about this kind of petroleum law, common in Africa today, which "nationalizes" oil and gas reserves, is that it takes ownership and possession away from the communities who live above the resources. Nationalization, by law, expropriates the oil and gas from those people who live in Cabinda, in the name of another people, the Angolan, to whom they have been forcibly annexed. This kind of appropriation by a central government produces an "indigenization discourse" (Ukiwo 2008) which is found elsewhere in Africa, such as the Niger Delta, Southern Sudan or Darfur. Looking up from the bottom, the "national" oil companies appear as the collaborators of foreign oil companies. FLEC was fighting for the "indigenization" of oil, trying to become a tiny, oil-rich state in the Gulf of Guinea, such as Equatorial Guinea, or São Tomé.

But one year after independence, in 1976, MPLA founded a new national oil company, *Sociedade Naçionao de Combustiveis* (Sonangol) as the "exclusive concessionaire" for oil exploration and development. Law N° 13/78 is the legal framework of Angolan collaboration with foreign oil multinationals. The Ministry of Petroleum might be the MPLA organ that is responsible for overseeing the industry, approving any exploration or production, regulating oil field production levels and gas flaring, and setting oil-tax rates. But Sonangol is the legal concessionaire over exploration, production, refinement, and distribution.

Law N° 13/78 authorizes Sonangol to enter *joint ventures* and *production-sharing agreements* with foreign oil companies, two very different kinds of business collaboration. In a joint venture, the concessionaire (in this case Sonangol) invites other corporations to split investment costs, in exchange for which it cedes a share of the crude oil produced that is equivalent to the share of investment made by these partners. If Chevron pays 50 percent of investments, for example, it receives 50 percent of the oil. In a production-sharing agreement, foreign companies bid for a contract, pay all of the costs, and make all of the investments. When production starts, the crude oil is contractually divided (on paper) into: (1) "royalty oil," which goes to the owner of the land (i.e. paid to the state); (2) "cost oil," which is paid to the contracting firms at a rate determined by their contract; and (3) "profit oil," which is divided three ways between the foreign oil companies, the national oil company, and the government.

All of Angola's oil concessions are production-sharing agreements *except Cabinda "Block 0,"* the massive oilfield just offshore, where

Chevron's expatriate workers fly by helicopter. Sonangol has a minority holding (41 percent) in this joint-venture concession with ChevronTexaco (39.2 percent) and its co-concessionaires, French Elf (i.e. Total, 10 percent) and Italian "Agip" (9.8 percent). (www. sonangol.co.ao) This Cabindan exception results from its concessions being negotiated during the 1960s, when joint ventures were the norm. Production-sharing agreements became common only after the nationalizations of the 1970s. Production-sharing agreements are more advantageous to the state, which is paid "royalty oil" and "profit oil" without bearing the costs of producing either.

But to understand why Angola has left Gulf with the concession in "Block 0," it is important to understand the illicit financing. Under the various oil contracts signed with the Americans, French, and Italians in "Block 0," a proportion of the money due to the Angolan state and Sonangol can be paid in crude oil. So a certain number of oil shipments belong to the state, a certain number belong to Sonangol, and the rest belong to Chevron, Elf, and Agip. This financial arrangement has allowed Sonangol to use a proportion of the national crude oil output to arrange a set of deals through overseas "escrow accounts." The company that receives a Sonangol cargo, in this case, has a choice. It can either sell it to a refinery, or sell it to somebody else. After it sells the oil, then it places that money paid into an account called an "escrow account." Under the lending agreement repayments for the loan come directly out of this account. Because much of the money flowing into escrow accounts is overseas, it does not go through the Angolan National Bank. Escrow accounts have become a set of parallel finances, where a large part of the national budget is effectively run extra-territorially through a set of foreign bank accounts.

Almost all of Angola's oil concessions are offshore, where the government has signed production-sharing agreements. At the signing of such major contracts, it is customary to pay a lump sum of money called a "signature bonus." These are substantial, one-off, non-recoverable down payments made in order to secure exploration and extraction rights from offshore blocks. Because they are paid before any oil is produced, they are not included in the normal calculations of oil revenues. These are the veritable wages of illicit collaboration. According to Nobel Prize-winning Global Witness (1999), high corruption of MPLA officials in the government and its national oil champion has resulted in the deployment of signature bonuses, outside the state budget, into the same parallel system of financing used in the escrow accounts: "Corruption starts with the head of state [Eduardo Dos Santos] surrounded by a clique of politicians and business cronies, collectively known as the *Futungo*, named after the Presidential Palace." (Global Witness 1999)

Of particular importance are "oil-backed loans," which account for a third of Angolan debt. With future oil output pledged to pay back these loans, new loans are regularly sought. To guarantee that the loans are repaid, foreign banks have arranged some offshore accounts as trust systems into which oil receipts can be paid directly. The oil leaves offshore enclaves, and the money goes into offshore accounts. Because the money does not go through the Angolan National Bank, and Sonangol discounts its repayments into these loans from the regular taxes it owes to the treasury, another set of parallel finances has emerged from oil-backed loans. This led Global Witness (1999) to describe the triumvirate of Sonangol, BNA, and *Futungo* as a "Bermuda triangle," where the money disappears offshore. The revenue that Sonangol receives from joint ventures is not reported, and the price of oil is underestimated in the state budget. Government expenditure statements are not accurate. According to the Economist Intelligence Unit, the system of parallel financing in Angola involves the large-scale diversion of revenue from the oil sector, with the result that much state revenue never enters the Treasury. The exact amount of money is not known.

Sonangol has become what Gary and Karl (2003) called a "state within a state," expanding into other sectors of the economy (mobile phone, airlines, insurance, shipping), effectively crowding out the private sector and giving it enormous leverage in what amounts to a monopoly in some sectors of the local market: "According to an IMF report, the company has never been independently audited, so there is no way to assess its performance." (p. 33)

The growing experience of Sonangol, and the greed for increased revenues, has contributed to an effort by the government to drive harder bargains with multinational oil companies. Angola has captured between 40 and 75 percent of revenues raised from its oil offshore. It has provided other African oil-producing countries with a model of commercial and political success. With fairly scarce human resources and a similar domestic setting to that of other African producers, it has developed into a credible commercial entity in the course of the last three decades. Following sustained problems with oil traders in the early 1980s, it established an operation in London that markets Angola's own share of oil, thus cutting out multinational middlemen. Sonangol consultants have been dispatched to Equatorial Guinea, São Tomé, Congo (Brazzaville), and Gabon. The image of Sonangol is one of success. We are often reminded that President Eduardo Dos Santos was a Soviet-trained petroleum engineer. But it is important to recognize that his collaboration and corruption, largely predicated on the greed for oil revenues, has kept enormous sums of money (estimated at some one billion dollars a year, perhaps more) from being re-invested in education of Angolan engineers, creation of Angolan offshore petroleum technology, or any of the other necessary investments in human capital to allow Sonangol to rid itself of foreign corporations. Therefore, so long as Angola continues on this path, foreigners will own and manage the oil of Cabinda.

3
International Organization and Governance

IMPROVING GOVERNANCE IN AFRICA'S OIL SECTOR

What motivates leaders to divert public oil revenues into private bank accounts? What are the real causes of political corruption? Before we can ever fight corruption, we must identify its causes. Political scientists such as Samuel Huntington and Joseph Nye, working on "modernization and development" theory, once argued that corruption was cultural, and played a positive role in developing countries. Obviously if this were the case, nothing should be done to fight corruption. Since then "new institutional economics" has been used to study corruption and has empirically demonstrated numerous negative effects of corruption on growth, investments, demand, costs, and progressive erosion of confidence in the state. (Kreuger 1974; Hutchcroft 1997; Heywood 1997) So we must fight corruption; and to do that, we must find its causes.

According to Collier and Hoeffler (2001) the root cause of corruption is "greed." A simple idea, grounded in human nature, greed is a psychological motivation pervasive in the literature on African oil. Greed is at the very center of the debate, for example, about oil and war (see Chapter 8). But what can be done to remove greed from people's hearts? Blaming human nature provides no clear course of action. Greed theory places too much emphasis on *individual* morality and not enough on the *institutions* by which they govern. Twenty years ago, Robert Klitgaard (1988) published a three-part formula explaining the institutional causes of corruption as a behavior that flourishes when government agents have (1) exclusive power over something, (2) a wide margin of discretionary power, and (3) are not held accountable for their acts. Corruption, therefore, is a consequence of bad institutional design. This is an example of how new institutional economics accepts human nature as it is, and focuses its attention instead on the way society's institutions affect performance. Bad institutions create micro-economic inducements for corruption. Good institutions reduce them. Strong bureaucracy, rule of law, and other elements of "good governance," according to this approach, can control corruption. But in order to reduce corruption one has to be able to measure it. How are "corruption" and "governance" measured? Transparency International publishes a *Corruption Perception Index* (CPI) that rank-orders countries on a ten-point scale from 0 ("highly corrupt") to 10 ("highly clean"). All African oil exporters have low CPI rankings: Gabon 3.3, Cameroon 2.4, Angola 2.2, Nigeria 2.2, Congo 2.1, Equatorial Guinea 1.9, Chad 1.8, and Sudan 1.8. But CPI is not a measure of corruption itself. It is an indirect measure of the *perception* of corruption. Some activists demand that oil companies and governments report their *actual* payments and revenues. Any difference in the numbers reported represents a direct factual measure of corruption itself. The World Bank measures corruption as a subset of governance. This chapter will examine the World Bank indicators: (1) voice and accountability, (2) political stability, (3) government effectiveness, (4) regulatory quality, (5) rule of law, and (6) control of "corruption, (where "corruption" is treated as a sub-set of "governance").

What is *governance*? The World Bank defines it as the "traditions" and "institutions" by which authority is exercised. (Kaufman, Kraay and Mastruzzi 2008) Efforts to fight oil corruption in Africa are a special case of "*global governance*," which refers to the process by which governments, intergovernmental organizations, and non-governmental organizations "come together to establish global rules, norms, and standards" (Higgot and Ougaard 2002: 208). Its World Governance Indicators (WGI) is a statistical *compilation*, not a direct observation. One common misconception is that the World Bank directly gathers the data itself. World Bank economists compile reports published by other global survey research institutes, think-tanks, non-governmental organizations, and intergovernmental organizations. In 1996 they started with 173 countries and 13 data sources. Today they cover 212 countries and territories from 35 independent data sources, representing the largest quantified governance database in the world.

Quantification means setting up a standard amount of a thing and putting a label on it. When economists do this, they make it possible to express qualitative concepts in a way that provides a common reference for observation. This is a useful approach when making comparisons. Numerous global organizations now provide annual rankings of governance and corruption, which it is possible to gather and arrange *longitudinally*, and then observe positive or negative *changes over time*. This measurement of change is straightforward: Either there is change in the qualitative state of governance, or there is not. Either governance is getting "better," or it is getting "worse." This chapter will compare WGI data for African oil-states. As the indicators are "ordinal" (i.e. rank-order) and few in number (N = 6), a simple yes/no–dichotomous measurement of change is considered valid methodologically (Menard 2002: 58).

WGI creators Kaufman, Kraay, and Mastruzzi (2008) are aware of the limitations of their methodology. "The composite indicators we construct are useful as a first tool for a broad cross-country comparison and for evaluating broad trends over time," they explain. But the "aggregate WGI are often too blunt a tool to be useful in formulating specific governance reforms in particular country contexts" (p. 6). For this, the authors suggest, case study is preferred, because it takes into consideration many other qualities and contextual factors shaping governance in an individual country. WGI is most useful when used in *comparative methodology*, providing theoretically informed measurements that can be used to compare increases or decreases in corruption across the cases. WGI can be used to evaluate the effectiveness of global anti-corruption initiatives through a simple bivariate analysis, i.e. measuring a relation between two variables. Governance being "better" (+) or "worse" (–) depends on the "presence" (+) or "absence" (–) of a global anti-corruption initiative. Thirteen years of WGI data (1996–2009) reveal positive and negative trends over time in the six qualities of governance.

Table 3.1 distinguishes between examples and counter-examples of *necessary* and *sufficient* conditions. A necessary condition is a *sine qua non*, without which the effect will not appear. A sufficient condition is always present before or at the same time as the effect it produces. To call something a cause of something else is to say it is both a necessary and sufficient condition. Only if *all* the cases are classified as necessary and/or sufficient conditions can we argue that oil-transparency initiatives are the "*cause*" of better governance. Since other factors are present and play a role, since there is rarely a perfect 100 percent association between two variables, a statistical measure called *phi* (Φ) is used to calculate a numerical *degree* of association between them. This is a conventional statistical measure of the strength of a relationship expressed by two or more sets of nominal or categorical data (i.e. "qualitative" data). Numerically, *phi* is a coefficient between –1 and +1, where positive one tells us that good governance is perfectly and directly associated with the oil initiatives, and where negative one implies a perfect inverse relationship. *Phi* allows us to speak in terms of degrees, rather than rigid categories. Additionally, it can be easily translated into a *Chi-square* test of statistical significance.

Table 3.1   Do oil initiatives cause better governance?

	Initiative Signed (+)	Not Signed (−)
Positive Change (+)	"good governance" Oil governance initiatives are *sufficient* for better governance.	"good governance" Oil governance initiatives are *not necessary* for better governance.
Negative Change (−)	"bad governance" Oil governance initiatives are *not sufficient* for better governance.	"bad governance" Oil governance initiatives are *necessary* for better governance.

International efforts to fight corruption in African oil-states today include a number of global governance initiatives to create transparency that are of particular importance. The method is to be able to identify where global governance initiatives are present, and where they are not, in order to thereafter measure their association with better or worse governance. First we shall identify and describe these transparency initiatives. Then we shall use the governance indicators to measure their effect on corruption.

INTERNATIONAL ORGANIZATIONS: TI, GLOBAL WITNESS, PWYP, EITI

Peter Eigen is the founder of Transparency International (TI). Most of his professional life he worked at the World Bank. He started as a legal adviser and eventually became the director of its regional commission for West Africa, East Africa and Latin America. For 25 years he was an eyewitness to corruption in those African countries where he was posted, such as Nigeria and Kenya. He also worked in Botswana, however, and was struck by how President Seretse Khama managed to turn his country into a prosperous nation by fighting corruption. Over time he became critical of the World Bank, where the subject of political corruption was taboo. His criticisms resulted in his dismissal by the Bank's director. "I knew that corruption was omnipresent," explained Eigen, "and that it hindered, if not disallowed, our development efforts." So in May of 1993 he founded Transparency International, a professionally staffed, global NGO whose stated mission was "to create change towards a world free of corruption." (www.transparency.org) TI claims to have played a leading role in improving the lives of millions around the world by building momentum for the anti-corruption movement.

TI is a global governance organization active in all four stages of the policy process. First, it raises awareness of corruption and the need for transparency (*agenda-setting*). Second, it advocates anti-corruption laws (*policy formulation*). Third, it lobbies for the application of anti-corruption laws (*policy implementation*). Fourth, it ranks and measures the changes in the perception of corruption over time (*policy evaluation*). It has been successful at getting transparency on to the global governance agendas of many important institutions. It has also been successful at formulating policy at the national level. The way TI works is by teaming up with governmental and non-governmental actors (often the former are the "whistle blowers") and gathering information on grand and petit corruption. It founds national chapters in each country, run by the citizens of that country. These are self-governing autonomous bodies that are confederated into a global network. Only the global headquarters are located in Berlin.

In 1993 it published a *National Integrity System*, an *Anti-Corruption Handbook*, and a *Corruption Fighter's Tool Kit*. These books teach ordinary citizens practical skills they need to fight corruption at the national level, such as how to analyze draft legislation, how to create integrity pacts, or how to monitor public institutions. In 1995 it established its now-famous *Corruption Perception Index* (CPI), a trademark product, and a useful tool for evaluating any positive or negative changes in corruption. On the policy formulation level it helped to draft new UN convention against corruption in 1996, and helped to create the *OECD Anti-Bribery Convention* in 1997. Two years later it created its *Bribe Payer's Index* (BPI) to get bribery on the global agenda. In 2003 it created a new *Global Corruption Barometer* (GCB) and also co-authored the *AU Convention on Preventing and Combating Corruption*, adopted by Nigeria in 2006. Everywhere it is a watchdog of corruption, acting at all levels, and making a name for itself in the process. Every three years it holds an *International Anti-Corruption Conference* (IACC), which brings together the makers and shakers of the anti-corruption movements in their country. At this conference TI gives transparency awards to brave and successful transparency activists, many posthumously.

Publish What You Pay (PWYP) is an international transparency initiative created by George Soros, a billionaire financier famous for running one of the world's biggest hedge funds. He was born in Budapest in 1930, survived the Nazi occupation and fled communist Hungary for England, where he graduated from the London School of Economics. He then settled in the United States, where he accumulated a large fortune through the investment advisory firm he founded and managed. He has been active as a philanthropist since 1979, and established the Open Society Institute, a network of philanthropic organizations that are now active in more than 50 countries dedicated to promoting the values of democracy and an open society. The foundation network spends about $400 million annually (www.georgesoros.com).

His really big idea of breaking the link between oil and corruption by requiring corporations to "publish what they pay" to the governments of oil-exporting countries had its intellectual origins in the writings of social activists Patrick Alley, Charmian Gooch, and Simon Taylor, the three co-founders of the London-based Global Witness in 1993. It was only upon reading their Nobel Prize-nominated report on oil and corruption, *A Crude Awakening* (1999), that Soros launched the Publish What You Pay campaign from his London headquarters of the Open Society Institute in 2002. Since then, PWYP has been joined by an "army of salvation." Founding members included the confederation of Transparency International, Oxfam, and Save the Children, joined by Catholic Relief Services, Human Rights Watch, and Secours Catholic, to name but a few of the morally credentialed institutional stakeholders. The moral force of these stakeholders is important, because there is little political support for mandatory disclosures of payments and revenues in either the oil industry or the regimes. National coalitions of PWYP are present in only three African oil-exporting countries: Cameroon, Congo, and Nigeria. (Gabon recently suspended its participation in 2008, and has arrested, imprisoned, and assaulted PWYP activists.) As Soros well realized, his call for *mandatory* disclosures is perceived as a dangerously radical position by both the companies and the regimes where oil money is the only game in town.

A less radical World Bank effort, the Extractive Industries Transparency Initiative (EITI), is a globally developed standard that promotes revenue transparency at the national level. It focuses on revenue resource transparency in African minerals-exporting countries through the following measures: increased accountability, reduced corruption, fostering democratic debate, improving macro-economic management, and improving access to finance. The first criteria enumerated in the *EITI Sourcebook* is similar to that of PWYP: the "regular publication of all material oil, gas and mining payments by companies to governments" (i.e. payments) and "all material revenues received by governments from oil, gas and mining companies" (i.e. revenues) to a wide audience in a publicly accessible, comprehensive and comprehensible manner (*EITI Sourcebook*, 2005). Its difference from PWYP is that EITI disclosures are *voluntary*. Four African oil-exporting countries have signed EITI (see Map 3.1).

The main author of EITI was former British Prime Minister Tony Blair who, in a speech at the World Summit in Johannesburg in 2002, told the delegates that because Africa was a personal passion for him (his father had been a missionary in Sierra Leone) he believed that "the world has a duty to heal the scars of the continent." Using his charisma to gather international support, Blair officially launched EITI in June 2003, and in December got the World Bank's endorsement. His personal diplomatic relations with George W. Bush persuaded the US president to endorse EITI at the Sea Island G8 World Summit in 2004. The next year at the Gleneagles G8 World Summit he established a Commission for Africa, whose goal was the implementation of EITI. What started as diplomacy has grown into a global governance initiative. In 2006 an International Secretariat of EITI was established in Oslo.

The way EITI works is by accepting countries as "candidates" into the organization. The metaphor is joining a club. Candidate countries are not coerced into joining, but must volunteer. This sets in motion a positive dynamic. Once they pass through a verification process (for example, once they have published their oil revenues sufficiently for the Secretariat) then they move to a "verified" category. Fifteen new "candidate" countries were welcomed in 2007, joined by eight more in 2008, making a total of 23 signatory states. Of the ten African countries in this study four are EITI candidate countries. Since the three presidents of Congo-Brazzaville, Equatorial Guinea, and Gabon are corrupt autocrats, however, their participation in EITI should be considered more as a diplomatic gesture toward foreign investors than a real commitment to revenue transparency.

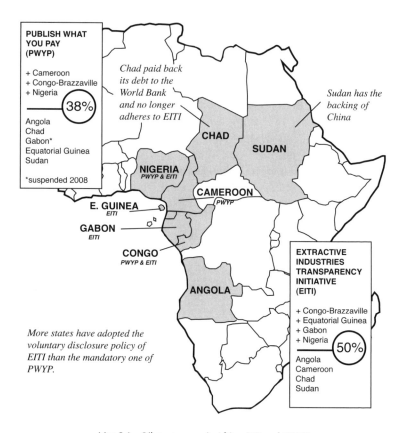

Map 3.1   Oil governance in Africa: EITI and PWYP

What relationship, if any, can we find between the presence of these global initiatives and the control of corruption? The World Bank defines the "Control of Corruption" (CC) as the measurement of the perceptions of the extent to which public power is exercised for private gain, including both petty and grand forms of corruption. (Kaufman et al. 2008: 7) Control of corruption also includes perceptions of the "capture" of the state by elites and private interests. Essentially, the private use of public office and resources is the core definition of corruption, and the need to liberate public institutions from the private control of political leaders and their "capture" by special interests is one of the greatest problems in establishing modern states in Africa. (Hellman, Jones, and Kaufman 2000) Table 3.2 shows on average a small negative change in the "control of corruption" in the African oil-states (−0.13).

Table 3.2 "Control of Corruption," –2.5 ("Very Poor") to +2.5 ("Very Good")

	(2005)	*(2006)*	*(2007)*	*(2008)*	*Change (+/–)*
Angola	–1.24	–1.21	–1.12	–1.22	–0.10
Cameroon	–1.15	–1.00	–0.93	–0.90	+0.03
Chad	–1.33	–1.20	–1.22	–1.45	–0.23
Congo	–1.04	–1.08	–1.04	–1.16	–0.12
Eq. Guinea	–1.53	–1.52	–1.37	–1.62	–0.25
Gabon	–0.66	–0.90	–0.85	–1.07	–0.22
Nigeria	–1.21	–1.14	–1.01	–0.92	+0.09
Sudan	–1.37	–1.15	–1.25	–1.49	–0.24
TOTAL	–9.53	–9.2	–8.79	–9.83	–1.04
AVERAGE	–1.19	–1.15	–1.10	–1.23	–0.13

Source: World Bank, *Governance Indicators* (2009)

"Regulatory Quality" (RQ) is the WGI measurement of the perceptions of the quality of the government to formulate and implement sound policies and regulations that permit and promote private sector development. (Kaufman et al. 2008: 7) *Good*-quality regulation involves mechanisms to estimate the costs and benefits of competing proposals, adequate technical and political information and a coordinated policy debate between the heads of the responsible agencies, feedback from other third-party stakeholders, professional monitoring and evaluation of implementation of the policy, competitive tendering of projects, performance-based benchmarks, decentralization of policy implementation, and opening up dialogue with advocacy groups from "civil society." *Bad* regulation is when officials collude with those they are meant to regulate, when enforcement becomes distorted over time, implemented so that corrupt behavior is not changed, or so that benefits fail to reach the intended beneficiaries. Bad regulation is heavily biased against the poor and weak. On average, the quality of regulation in the ten African oil-dependent states has slightly worsened (–0.01).

"Political Stability and the Absence of Violence/Terrorism" (PV) is the bank's measure of perceptions of the likelihood that the government will be destabilized or overthrown by unconstitutional or violent means, including politically motivated violence and terrorism. (Kaufman et al. 2008: 7) It is common sense that the proper performance and maintenance of state functions and institutions provides the best protection against state collapse. Sovereignty embodies a nation's ultimate self-determining powers over its own future, which is entrusted in conditional custody by its people to the state. But African states have had great difficulty in maintaining sovereignty within their borders. This is sometimes because of the power of local, or private, or regional bosses. Sometimes it is because the state's legitimacy is challenged by groups who don't want to be a part of it. Sometimes it is because of political opponents who refuse to accept the incumbent regime. Military coups, violent rebellions, and terrorist attacks all contribute to government concentrating its resources on short-term survival, rather than on long-term policies. Political instability creates a breeding ground for corruption. According to the WGI data, African oil-dependent states, on average, have poor ratings of political stability (−0.98) though slightly improving (+0.02) (see Table 3.4).

Table 3.3   "Regulatory Quality," −2.5 ("Very Poor") to +2.5 ("Very Good")

	(2005)	*(2006)*	*(2007)*	*(2008)*	*Change (+/−)*
Angola	−1.35	−1.31	−1.08	−1.30	−0.22
Cameroon	−0.72	−0.73	−0.71	−0.66	+0.05
Chad	−1.06	−1.09	−1.16	−1.26	−0.10
Congo	−1.23	−1.07	−1.20	−1.19	+0.01
Eq. Guinea	−1.34	−1.34	−1.35	−1.37	−0.02
Gabon	−0.34	−0.49	−0.71	−0.65	+0.06
Nigeria	−0.89	−0.96	−0.89	−0.62	+0.27
Sudan	−1.25	−1.16	−1.25	−1.36	−0.11
TOTAL	−8.18	−8.15	−8.35	−8.41	−0.06
AVERAGE	−1.02	−1.02	−1.04	−1.05	−0.01

Source: World Bank, *Governance Indicators* (2009)

Table 3.4   "Political Stability," −2.5 ("Very Poor") to +2.5 ("Very Good")

	(2005)	*(2006)*	*(2007)*	*(2008)*	*Change (+/−)*
Angola	−0.78	−0.44	−0.46	−0.43	+0.03
Cameroon	−0.39	−0.31	−0.39	−0.53	−0.14
Chad	−1.32	−1.87	−1.96	−1.92	+0.04
Congo	−1.23	−0.97	−0.83	−0.61	+0.22
Eq. Guinea	−0.33	−0.09	−0.16	−0.09	+0.17
Gabon	+0.08	+0.13	+0.20	+0.23	+0.03
Nigeria	−1.73	−2.05	−2.07	−2.01	+0.06
Sudan	−2.12	−2.13	−2.30	−2.44	−0.14
TOTAL	−7.82	−7.73	−7.97	−7.80	+0.17
AVERAGE	−0.98	−0.97	−1.00	−0.98	+0.02

Source: World Bank, *Governance Indicators* (2009)

"Government Effectiveness" (GE) is the bank's measure of perceptions of the quality of public services. This includes the quality of the civil service, and the degree of its independence from political pressures, the quality of government policy formulation and implementation, and the credibility of the government's commitment to such policies. (Kaufman et al. 2008: 7) A crucial requirement for government effectiveness is a unified and competent bureaucracy, based on merit recruitment and offering stable and rewarding careers relatively free from political interference by sectional interests that might compromise the pursuit of economic development. Effective government is where public offices belong to the state, not to the office-holders personally. Occupancy of such office should entail no powers of private patronage nor be used for the support of any particular private client base (no clientism). Absence of a proper bureaucracy allows corruption to flourish, and erodes the legitimacy of the state. Officials respond to corrupt incentives to create scarcity, delay, and red tape. Public resources are misallocated, revenue is lost, and public service morale is undermined. In Africa, neo-patrimonial recruitment and promotion is commonplace, leading to trust and confidence in the state evaporating. The basic characteristic of clientism is unequal power relations between a patron (big man) who is high-status, powerful, and rich, on the one hand, and a client (small boy) who lacks power, wealth, or status, on the other hand. Their relationship is reciprocal, but uneven, in that the patron has control of resources and opportunities he can provide to his clients in return for deference, support, loyalty, and votes. Table 3.5 shows that, on average, government effectiveness perceived as "poor," and has declined slightly (–0.02) in recent years in the African oil-exporting states.

Table 3.5   "Government Effectiveness," –2.5 ("Very Poor") to +2.5 ("Very Good")

	(2005)	(2006)	(2007)	(2008)	Change (+/–)
Angola	–1.01	–1.23	–1.16	–0.98	+0.18
Cameroon	–0.93	–0.84	–0.87	–0.80	+0.07
Chad	–1.20	–1.32	–1.45	–1.48	–0.03
Congo	–1.35	–1.29	–1.34	–1.34	0.00
Eq. Guinea	–1.36	–1.34	–1.37	–1.43	–0.06
Gabon	–0.71	–0.69	–0.66	–0.70	–0.04
Nigeria	–0.84	–0.89	–0.93	–0.98	–0.05
Sudan	–1.48	–1.12	–1.18	–1.41	–0.23
TOTAL	–8.88	–8.72	–8.96	–9.12	–0.16
AVERAGE	–1.11	–1.09	–1.12	–1.14	–0.02

Source: World Bank, *Governance Indicators* (2009)

"Voice and Accountability" (VA) is the bank's measure of perceptions of the extent to which a country's citizens are able to participate in selecting their own government, as well as the freedom of expression, the freedom of association, and the existence of a free media. (Kaufman et al. 2008: 7) Democratic, party-based government offers a responsible approach to economic management, because a long-run reputation for being responsible has an effect on a party's political fortunes. It handles domestic conflicts better than authoritarianism, and its accountability mechanism makes the gross abuse of public resources less likely. Bad governance is where an autocratic state is led by a rapacious officialdom, a "predatory state" that exploits the people for the benefit of the rulers, and holds back on development. When neo-patrimonial government is predatory, its political legitimacy is weak. So power usually has to be secured by the exploitation of ethnic loyalties, by patronage, by coercion, and by repression of the opposition. Table 3.6 shows that, on average, the quality of voice and accountability is poor, and has slightly declined in African oil states (–0.03).

Table 3.6   "Voice and Accountability," –2.5 ("Very Poor") to +2.5 ("Very Good")

	(2005)	*(2006)*	*(2007)*	*(2008)*	*Change (+/–)*
Angola	–1.21	–1.20	–1.11	–1.07	+0.04
Cameroon	–1.04	–0.96	–0.94	–1.02	–0.08
Chad	–1.40	–1.41	–1.43	–1.45	–0.02
Congo	–1.01	–1.06	–1.11	–1.16	–0.05
Eq. Guinea	–1.64	–1.84	–1.89	–1.89	0.00
Gabon	–0.86	–0.83	–0.83	–0.84	–0.01
Nigeria	–0.75	–0.49	–0.54	–0.60	–0.06
Sudan	–1.70	–1.74	–1.73	–1.77	–0.04
TOTAL	–9.61	–9.53	–9.58	–9.80	–0.22
AVERAGE	–1.20	–1.19	–1.20	–1.23	–0.03

Source: World Bank, *Governance Indicators* (2009)

Finally, "Rule of Law" (RL) is the bank's measure of the extent to which agents have confidence in, and abide by, the rules of society. In particular, the bank evaluates the quality of "contract" enforcement, "property" rights, the "police" and the "courts," as well as the likelihood of "crime" and "violence." (Kaufman et al. 2008: 7) Rule of Law is a measure of the degree to which the *institutional* rules of the game are followed by society. Institutions are more than organizations, although they may depend on organizations for their effectiveness. For example, a rule of law depends on an impartial court. Property rights depend on an institutional framework of legal rules, organizational forms, and norms of behavior. Improvement in the quality of institutions is correlated with increases in growth, the level of per capita income, economic growth, and stability. (Edison 2002) Failure of the state to organize its apparatus of power effectively and to create political order leads to ineffective public administration and corruption. Laws are not passed, order is not preserved, social cohesion decays, security (especially for the poor) is lost, and legitimacy evaporates.

Bad governance is the rule of men, not law. New institutional economics considers this quality more important to economic development than conventional quantities such as investment or wages. Investing in a country with poor legal institutions, *ceteris paribus*, is less conducive to development than investing in a country with good rule of law. Table 3.7 shows, on average, that the respect for the rule of law in African oil states has not changed (0.00) over recent years. Rather it shows that, on average, they are ranked "very poor" (−1.19), and the respect for law in African oil states remains more than one standard deviation below the mean.

Table 3.7   "Rule of Law," −2.5 ("Very Poor") to +2.5 ("Very Good")

	(2005)	*(2006)*	*(2007)*	*(2008)*	*Change (+/−)*
Angola	−1.40	−1.28	−1.35	−1.28	+0.05
Cameroon	−1.07	−1.03	−1.09	−0.99	−0.02
Chad	−1.33	−1.38	−1.40	−1.57	−0.07
Congo	−1.46	−1.24	−1.26	−1.16	+0.20
Eq. Guinea	−1.33	−1.24	−1.16	−1.31	+0.17
Gabon	−0.51	−0.64	−0.60	−0.62	−0.09
Nigeria	−1.41	−1.19	−1.20	−1.12	+0.21
Sudan	−1.61	−1.34	−1.46	−1.50	+0.15
TOTAL	−10.12	−9.34	−9.52	−9.55	−0.03
AVERAGE	−1.27	−1.17	−1.19	−1.19	0.00

Source: World Bank, *Governance Indicators* (2009)

The method of statistical inference is fairly simple. Take the two variables of interest, transparency initiatives, and better governance, and score each country as either positive (+) or negative (−) on those two qualities. This produces four possible outcomes (+/−, −/+, +/+, −/−), which can be tabulated on a 2x2-contingency table (Table 3.8). The frequencies of each cell are counted, and input into the formula of the phi coefficient.

Table 3.8   Effects of transparency initiatives on governance (Φ = 0.60)
(*Necessary* but *Not Sufficient* for Better Governance)

	Initiative Present (+)	Initiative Absent (−)
Positive Change (+)	Nigeria, Gabon, Cameroon (3)	(0)
Negative Change (−)	Congo, E. Guinea (2)	Angola, Chad, Sudan (3)

Phi (0.60) suggests a 60 percent association, and Chi-square (2.88) suggests a 90 percent probability ($r \geq 0.90$) that the relationship between a transparency initiative and better governance is statistically significant (i.e. not due to chance). Also this analysis suggests that even if transparency has been *necessary* for better governance, it has *not* been *sufficient*. For if Nigeria, Gabon, and Cameroon are three positive examples of EITI promoting better governance over time; and if Angola, Chad, and Sudan are three negative examples suggesting transparency is necessary; two counter-examples (Equatorial Guinea and Congo) are cases of EITI signatories with worsening governance indicators. So the World Bank data suggest that EITI has been *insufficient in fighting corruption*.

CASE STUDY: CHAD AND THE WORLD BANK MODEL

Chad is the single clearest failure of EITI. It is a case study in how *not* to implement good governance in the petroleum sector. Ironically, Chad was the showpiece of what was once heralded as a "World Bank Model." Nobody talks about Chad as being a model of "good governance" in the African oil sector any more. Not only do critics berate its authoritarian dictator Idriss Déby as a corrupt praetorian who spends his oil revenues to cling violently to power. Using the World Bank's own data, in Table 3.9, we can see that since oil exports have started in 2003, all six of the WGI governance indicators have declined dramatically. This is very significant because the World Bank promoted the development of oil in Chad by promising that improvements in governance could be produced at the same time as oil. Therefore the World Bank had the most at stake in reporting better governance indicators.

Chad should not be considered a model (or anti-model) for African oil-exporting countries anyway, because several special circumstances make it an exceptional case. First, it is landlocked, so

Map 3.2   Chad–Cameroon pipeline

Table 3.9   Bad governance in Chad 1996–2007

	CC	RQ	PV	GE	RL	VA
1996		−0.86	−0.74	−0.65	−0.88	−0.90
1998	−1.00	−0.94	−1.31	−0.61	−0.97	−0.96
2000	−0.88	−0.80	−1.36	−0.62	−0.91	−0.97
2002	−0.93	−0.86	−1.60	−0.87	−0.80	−0.90
2003	−1.13	−0.94	−1.23	−0.83	−1.07	−1.05
2004	−1.17	−0.78	−1.22	−1.02	−1.14	−1.20
2005	−1.33	−1.06	−1.32	−1.20	−1.33	−1.40
2006	−1.20	−1.09	−1.87	−1.32	−1.38	−1.41
2007	−1.22	−1.16	−1.96	−1.45	−1.40	−1.43
2008	−1.45	−1.26	−1.92	−1.48	−1.57	−1.45
(Avg.)	(−1.15)	(−0.97)	(−1.45)	(−1.00)	(−1.14)	(−1.17)

(CC) Control of Corruption, (RQ) Regulatory Quality, (PV) Political Stability & Absence of Violence, (GE) Government Effectiveness, (RL) Rule of Law, (VA) Democratic Voice and Accountability

Source: World Bank Governance Indicators (2009)

getting its oil to market required massive investment in a *pipeline.* Second, it is post-conditionally aid dependent and extremely poor, so World Bank *leverage* was unusually strong. Third, Chad is a "collapsed state" (Zartman 1995) barely recovering from civil war, so the oil companies needed the Bank's involvement to avoid the international *criticisms* then being levied against the analogous pipeline project in war-torn Southern Sudan. "This combination of factors," noted two activists from Catholic Relief Services "may not be seen again." (Gary and Karl 2003: 74) Perhaps the people of Niger can try to learn from the failure of Chad that any successful improvement in the governance of oil requires more than international pressure. Good oil governance requires transparency, political leadership, and collective action from below. Without these domestic–internal elements, no foreign "model" will alone suffice.

In other words, the story of the Chad–Cameroon pipeline is worth telling because it reveals the problems of trying to improve oil governance from above. Widely touted as a model for other oil-exporting countries, this $3.7 billion project brought together ExxonMobil, ChevronTexaco, Petronas (Malaysia), the World Bank, the governments of Chad and Cameroon, and "civil society" into a new set of *governance institutions* (laws, independent agencies, pacts with civil society, etc.) to use Chad's oil revenues for peace, poverty alleviation, and socio-economic development. According to documents published by the oil consortium, they planned to pump 1 billion barrels of oil from the Doba oil fields in southern Chad over 25 years, with a peak production of 225,000 barrels per day in 2004, gradually declining to 150,000 bpd by 2009, and 100,000 by 2013 (ExxonMobil 1998). But activists pointed out that once Doba fields were online, other oil discoveries in southern Chad (estimated reserves) could be connected to the pipeline at the fields around Doba (proved reserves).

Construction of the pipeline was started in 2000 and was completed by 2003. I visited the pipeline myself in late 2002, just before the oil started flowing. It looked like a straight line cut through the forest (the pipe was buried underground, "invisible"), like a corridor of flattened forest that sliced a line straight down to the coastal town of Kribi, Cameroon, where the crude was connected by an underwater pipeline to an offshore tanker terminal. This pipeline symbolized the many problems of the African oil sector: foreign domination, export orientation, environmental pollution and grievances of indigenous peoples, rent-seeking, political corruption, and violent conflict. It also told an important story about the hubris of foreign actors, in this case World Bank economists, who think they have all the answers. The Doba oilfields would not have been developed without them. World Bank involvement triggered easy access to export credit agencies, and private bank financing. It provided a "political cover" for the oil consortium to build a pipeline under circumstances analogous to those of war-torn Sudan. Perhaps more than any other player, the World Bank took a huge reputational gamble by claiming that its assistance would promote the good government policies necessary to avoid the resource curse.

All the country experts recognized a real possibility of its failure from the very start. The risks were high, because Chad was the prototypical case study of a "collapsed state." Readers should be reminded that in those days Chad was famous as the very first chapter of I. William Zartman's seminal book *Collapsed States* (1995), which coined the term. Analysts, scholars, and educated people at the World Bank, most certainly knew that Chad had suffered from a large number of revolts, rebellions, assassinations, extra-judicial killings, coups d'état, foreign military interventions, regional successions, and a civil war that touched every corner of the country: "No part of the country escaped armed violence; no Chadian family escaped the violence unscathed." (Foltz 1995: 15) Several explanations have been given for the state failure and collapse in Chad. First, it has one of the most *ethnically diverse* social mosaics in Africa. Foltz (1995) counted between 72 and 110 different language groups. The CIA *World Factbook* estimates over 200 ethnicities. These ethnic groups have fractionalized into highly segmented politico-guerrilla groups where "bloody fights *between* fractions of the same ethnic group" were more common than conflicts "in which ethnic groups confronted one another as *blocs*." (Foltz 1995: 17).

Playing on this ethnic division, Muammar al-Gaddafi of Libya crossed the northern border at the Aouzou Strip and annexed one-third of Chad's territory in the chaos. This raises the second major explanation for the state collapse. Chad had *no natural borders* and six neighbors (Libya, Sudan, Nigeria, Niger, Cameroon, and the Central African Republic) so it required "hard" military and "soft" diplomatic power to keep the lines on the map colonialism had drawn. Finally, the colonial legacy of France had done little to build state capacity. Chad was "France's Cinderella colony" (Buijtenhuijs 1989: 54), utterly neglected economically and educationally. Chad suffered from a dramatic lack of well-trained civil servants at the beginning of the 1960s, people to man the state apparatus, especially at the regional and local level. After decolonization France had used Chad as a neocolonial battlefield in its regional struggle against Libya for mastery of the Sahel, and civil war led to the collapse of the state. All the government buildings in N'djamena were sacked and pillaged. All government functionaries eventually fled the capital city for their lives. The last government salaries were paid in August 1979, and state authority definitively "collapsed" in 1980.

Reconstruction started in 1982, when Hissein Habré's *Forces Armées du Nord* (FAN) took the capital from a weak transitional government, a hydra of ethno-political factions. Foltz (1995) claims that Habré managed to accomplish five basic elements of state reconstruction in his eight years of rule, before he was overthrown in a coup d'état in November 1990. Habré reconstructed *central political authority*. (He was a tyrant, but that is only proof that he had reconstructed central power.) Habré also re-established state control over his country's 5,968 km of *national boundaries*, preventing penetration by neighboring states to plunder its natural resources. Habré established a sufficient level of order over the *national territory* to prevent violent armed challenges to state authority. He established the state capacity to *extract resources* sufficient for the regime to function and reproduce itself. Finally, he controlled the *actions of his state agents* sufficiently to coordinate and execute policy. The capacity of the state to organize its apparatus of power effectively and so create political order is a necessary precondition for good governance. Nobody has written poetry about the beauty of the Habré regime, but he left behind a sovereign state to his successor, Idriss Déby.

This is the republic that the World Bank was going to try and improve in the 1990s: a violent, ethno-political military dictatorship. Because the Bank had justified its involvement to critics by promoting the poverty alleviation potential of the project, it forced Déby into agreeing to these and other measures, and even provided him with a $41 million loan to develop a revenue management and financial control system. This was an exorbitant sum, to be paid back with future oil revenues. The December 30, 1998, "Petroleum Revenue Management Law," *Loi No. 001/PR/99*, was voted unanimously by the National Assembly—after only three hours of debate! This suggests that its text had not been drafted or even debated in the parliament, and that Idriss Déby was not its true author. Rather, he was its Achilles heel.

The revenue management law (1999) was a novel institution of *global* governance, dividing "direct" oil revenues (i.e. royalties and dividends) and depositing them into offshore accounts with international banks, where the money was divided into two parts: (1) 10 percent deposited in a Norwegian-style "Fund for the Future," a long-term investment portfolio aimed at poverty reduction of future generations who will live in a post-oil future; (2) 90 percent deposited in the Chadian Treasury for the current generations. Out of that 90 percent, the money was earmarked into three parts: 80 percent for education, health, infrastructure, rural development, and environment, 15 percent for government expenditures, and 5 percent for the Doba region. (www.ccsrp. td) The law also established a Petroleum Revenue Oversight and Control Committee, or *Collège de contrôle et de Surveillance des Ressources Pétrolieres*, the CCSRP. The *Collège* was an independent government–civil society committee whose task was to verify, authorize, and oversee expenditures of oil revenues. "Everything depends on whether the law will be adhered to," admitted the US Ambassador in N'djamena to Ian Gary a year before the oil started flowing, reflecting the low perception of confidence in "Rule of Law" in Chad.

As for "Regulatory Quality," it might be noted that the World Bank did not undertake an economic cost–benefit analysis of alternatives to the pipeline project, and there were several other problems with the new law's formulation. It did not, for example, include indirect oil revenues (i.e. income taxes and customs duties) in its scope, estimated to generate almost half of total revenues generated by the project over its 25-year lifespan. The law only covered the three oil fields in Doba, and so it did not include any other oil fields in southern Chad, with estimated reserves equal to or greater than those of Doba. The law's small payment of 5 percent to local communities was arbitrary and inadequate to remedy the social, environmental, and property grievances of villages along the pipeline route. The law's social and economic spending was so vaguely defined that it did not distinguish between building a luxury hospital in the capital from building a rural hospital in the desert. The law did not set aside any portion of the oil revenues in an "oil stabilization fund," which deposits oil windfall profits into an account to offset price volatility and falling oil prices. Finally, the law did nothing to strengthen the judiciary or any other branches of the central government to serve as a counterweight to presidential decrees. (Gary and Karl 2003: 73)

By the time the pipeline was finished in 2003 a series of grievances from local communities affected by the pipeline (and not just those entirely wiped off the map by the football-field-wide corridor cut through the forest) included a significant migration of people from other regions who spontaneously settled in their region. There were also complaints about excessive dust caused by the construction, and the contamination of water reservoirs by the underground burial of the pipe. Inflation in the prices of basic commodities and housing also occurred as foreign workers arrived in their villages. Located in the middle of an equatorial rainforest, oil-workers earned salaries that largely exceeded the entire income of these rural villages. Locals also complained about the long delays in the delivery of promised aid programs to help local entrepreneurs to sub-contract. Finally, and most poignant, were the complaints that the village schoolteachers were leaving their schools to take well-paying construction jobs on the pipeline. The oil consortium had to manage this public relations catastrophe by agreeing to devote significant resources to social and environmental "safeguards."

The first thing that ExxonMobil did to meet environmental concerns about oil spills and the possibility of bunkering (theft of crude oil) by gangs of disgruntled youths (as in the Niger Delta) was to bury the pipeline underground. Next the consortium hired 112 professional staff members to work in its environmental group, who processed 4,120 "compensation" claims from villagers who lived along the corridor. The consortium also paid 226 villages an additional "regional" compensation payment for overall "externalities." Furthermore, the consortium re-routed the pipeline from its initial path in such as way as to avoid "environmentally sensitive" areas and to protect "indigenous communities." One group that received special attention from this effort was the Pygmies. The government of Cameroon even created two national parks in compensation for the environmental damage caused by the pipeline's construction. (Gary and Karl 2003: 65) But Chad did not.

On the contrary, the Déby regime had a less rosy plan for Doba than a gamepark for tourists. The World Bank management argued that administrative capacity could be built in Chad at the same time as the pipeline, rather than preceding the start of construction. But its own International Advisory Group doubted the ability to develop both at the same pace, and called it a two-speed problem: "The commercial project is moving forward, while the institutions are limping along." (Gary and Karl 2003: 65) One of the most revealing moves made by Déby was his sacking of Amine Ben Barka, the national director of the Central Bank of Equatorial Africa (BEAC) in Chad. Ben Barka was removed from his post with no explanation.

But it is not hard to find one. Ben Barka had been the president of the Revenue Oversight Committee, one of the new governance institutions created by the petroleum revenue management law. He was seen as a competent, independent, technically skilled, and widely respected administrator. Since his membership on the Committee was dependent on his holding *ex officio* the Central Bank post at BEAC, his removal from the BEAC, and thus from the Committee, on the eve of Chad"s first oil export revenues, was a blow to "Effective Government."

The Constitution promulgated by Hissein Habré ostensibly created an independent judiciary with an elected High Court of Justice. The laws clearly stated that High Court Justices would be "elected" to their offices; and the government would "provide" for their election. This was intended to make the judiciary independent from the executive and legislative. But in May 2000, President Déby and the National Assembly *appointed* 15 members of the High Court of Justice, in strict violation of the constitutional separation of powers. In 2002 the State Department reported that while the Constitution mandates an independent judiciary, "the judiciary is ineffective, underfunded, overburdened, and subject to executive interference." (Gary and Karl 2003: 97) In its project appraisal document on the pipeline the World Bank claimed, contrary to the evidence, that"Chad has successfully put in place democratic political institutions" (World Bank 2000) but Chad's republican institutions were flagrantly violated by presidential (2001, 2006) and legislative (2002, 2006) elections which even the US State Department has reported to be "fraudulent," with "widespread vote rigging" and "local irregularities," that permitted President Déby to later amend the Constitution so as to remove the term limits on his office, and to hold an unbeatable majority of seats in the National Assembly. Only turning a completely blind eye to this spectacle of electoral authoritarianism allowed the World Bank to claim that Chad was improving its democratic "Voice and Accountability."

This brings us to the "Control of Corruption." The cause of oil corruption is no great mystery, but a collection of institutional incentives and inducements, where government agents are not held accountable for their acts, they have wide discretionary power, and they have exclusive power over the oil sector. This opportunity to corrupt is institutional; the choice to be corrupt is human nature. The first evidence of oil corruption in Chad came in late 2000, three years before the first drop of oil was exported, when the government announced that it had spent the first $4.5 million of a $25 million signature bonus on military weapons. The revenue management law did not technically cover such signatory bonuses, which were paid by the oil consortium whenever a government signed a contract. But clearly this was a violation of everything the World Bank had promised. Chad had known 30 years of bloody civil war. Its government at that time was fighting rebellions against rebel Zaghawa clansmen in the East (enraged that President Déby was not doing enough to protect his ethnic kinsmen from massacres in Darfur). Furthermore, a potential conflict was brewing with Sudan itself. As Doba reached peak oil production, Déby found himself in need of more weapons to fight off no less than three separate coup attempts coming from the East (2007, 2008, and 2009).

Of course, the World Bank had to continually paint a "rosy picture" of the violence that was ever-present in Chad. While fighting was going on in the border with the Central African Republic in September 2002, for example, a World Bank officer in Cameroon assured activists in his office in Yaoundé that "Chad is now at peace." (Gary and Karl 2003: 64) This was in complete contradiction of reliable information coming from civil society. *Amnesty International* regularly drew the attention of international opinion to the systematic use of summary and extra-judicial execution against unarmed civilians. Hundreds of civilians had been killed, for example, when Déby gradually consolidated his power at Ouaddaï (1994), Logone Occidental (1994–95), and Logone Oriental (1994–95) "and not one perpetrator of these violations has been brought to justice." (Gary 2003 Karl: 97) *Africa Confidential* reported on the scandalous violence against Ngarlejy Yorongar, a member of parliament, an active opposition leader, and a presidential candidate, who was arrested, beaten, and tortured in 2001. The World Bank president personally intervened to help free Yorongar from prison, and later members of the bank's Inspection Panel visited this opposition leader in a Paris hospital where they saw scars of his previous torture (August 30, 2002: cf. Gary and Karl 2003: 99).

Surely good governance does not mean that international financial institutions perform all of the good governance, and domestic government performs all of the bad. "If the history of development assistance teaches us anything," admitted the World Bank, "it is that external support can achieve little where the domestic will to reform is lacking." (World Bank 1999) Déby paid back his loan to the World Bank, and suspended its participation in EITI in 2009.

4
Rentier States and Kleptocracy

This chapter makes the argument that the oil states in Africa are *rentier states*: a typological category first coined by the Iranian economist Hossein Mahdavy (1970), writing about the problems of oil dependency in Iran. His theory was successfully applied by scholars working on the Arab world (e.g. Beblwai and Luciani 1987), Africa (e.g. Yates 1996), and Latin America (e.g. Buxton 2008; Campodócino 2008). More than a simple pejorative, this classification of "rentier state" really refers to a complex of associated ideas concerning negative developmental patterns in economies dominated by external rent, particularly oil rent. The first part of this chapter will define "rent" (Malthus 1815), "rentier," (Ricardo 1821), and the meaning of the term "rentier state." (Mahdavy 1970) It will also explain how this type of state is formed by a rentier economy (Beblawi 1987), which also gives rise to a tiny "rentier class," driven by a "rentier mentality." (Beblawi and Luciani 1987)

The second part of this chapter applies this template to the case of Equatorial Guinea, a country blessed with a small population, large oil revenues, and high rates of growth in GDP. But, paradoxically, it suffers from high unemployment rates, limited diversification, few job opportunities, marginal investment in local business, enclave industries with foreign workers, and shockingly high poverty rates—especially for a country reputed to be so "rich" in oil! This is called the "paradox of plenty" (Karl 1997; Gary and Karl 2003), the "resource curse" (Auty 1993), here explained by the logic of economic rent.

OIL RENT AND THE RENTIER

What is "*rent?*" In classical economic theory, it was understood to mean any surplus that was left over after all the costs of production had been met. It was paid to the owner of the land. You can think of it as an *income* paid to a landlord for the value of his real property, and for interference with his possession of land. An old English term of Norman French origins, feudal *rente* survived long enough to enter into the lexicon of nineteenth-century political economy. Rent is one of four factor incomes. The other three are "wages," "interest," and "profit." Thomas Malthus (1815) defined rent as "that portion of the value of the whole produce which remains to the owner of the land." David Ricardo (1821) defined it as a "gift of nature" paid to the owner for the "scarcity" of land, and its "difference in quality." Farmers pay rent to a landlord from what they get in price on the market and what it cost them to get it there in the first place (e.g. cost of seeds, tools, wages, interest, profits, merchants, etc.). If all these costs equaled the sale price, as might happen on marginally fertile land, the crop produced no rent. Cultivation of higher-quality land, on the contrary, could result in rent. Ricardo called this kind of income "differential rent" because it reflected the difference in the land's fertility, and so was a gift of nature.

Mineral rents also derive from the difference between the mineral price and the costs of production. "Mines, as well as land, generally pay rent to their owners, and this rent is the effect and never the cause of the high value of their produce." (Ricardo 1821: 590) Mineral rents reflect both the *quality* of land and the *market price* for the ore. Rent is paid to the owner of land for use of the land and removal of its resources. Since some lands are better in quality, their mineral deposits are richer or are better located than others, these lands tend to produce rent. "The general rule is that deposits where rents are greatest are the one most profitable to exploit now, and are determined by the price of the mineral on the present and projected future markets." (Pierce 1986: 99)

"Oil rent" can be defined as the difference between the price of a given quantity of oil sold to consumers in the form of petroleum products and the total cost incurred in discovering, producing, transporting, refining, and marketing that oil. The term as it is used here is not familiar to most people. Newspapers write about "oil windfall profits." That blurs a very real distinction between profits and rents. Also micro-economists use the term in a way that divorces it from the larger social context. Treating rent as a micro-economic category, something to be paid in a particular place and time, they tell us very little about the "rentier" as a larger social actor, as a member of a group or class who does not participate in the productive process, but still receives an income. A rentier is unlike any other social actor in the production process. The laborer receives wages for his work, the creditor receives interest for scarcity of capital, and the entrepreneur only receives profits when he successfully manages risk (each income derives from some element of sacrifice or risk). But the rentier receives his income "unearned," an entitlement of land ownership.

The term "Rentier state" was first used by Mahdavy (1970) to designate *any country that receives on a regular basis substantial amounts of external economic rent* (p. 428). The quantitative threshold was not specified, but Mahdavy was talking about cases where the effects of the oil sector on that country were "significant" (essentially oil-dependent countries). When he first coined the term, oil prices were around $3 a barrel, and so the oil rents he was talking about were still relatively small, by current standards. Mahdavy cited Kuwait and Qatar as extreme examples, with limited capabilities for industrialization and few alternative sources of revenue. But he was ahead of his time. His idea assumed greater relevance in the 1970s. The OPEC oil cartel, the Arab oil embargo, and the fall of the Shah of Iran combined to push the price of oil (thus rents) to historic highs: Oil prices rose tenfold, from around $3/bl (1970) to around $35/bl (1980) with no additional costs in production. Much of the oil price consisted of rent paid by companies (and ultimately consumers) to the producer states. "Typically," reported Luciani a decade later, "rents comprise 95 to 97 percent of gross receipts of low-cost oil." (Beblawi and Luciani 1987: 26)

A "rentier economy" has four preconditions identified by Hazem Beblawi (1987):

1. The economy must be one where rent situations predominate, *where more than 40 percent* of national income is derived from *oil*.
2. The origin of rent must be "external" to the economy, that is, from *foreign sources*. Domestic rents, even if substantial enough to predominate, are not sufficient to characterize an economy as rentier because they result from domestic factors of production. (If an economy is producing rent domestically, then it must also be producing domestic wages, interest and profit.)
3. *Only the few receive rent* in a rentier economy. An open economy with high levels of foreign trade is not rentier, even if it predominantly depends on rent, because domestic society is actively involved in the accumulation of wealth.
4. *Government must be the principal recipient* of the rent. This last characteristic is closely related to the concentration of rent in the hands of the few. "The state or the government, being the principal rentier in the economy, plays the crucial role of

the prime mover of economic activity." (Beblawi and Luciani: 53) The bureaucracy has a tendency to turn into a rentier class. (Mahdavy 1970: 467)

"Rentier space" is a recent invention of Kenneth Omeje (2006) to expand the number of factors and include the many complex contexts that have resulted in the violent Niger Delta conflict of Nigeria. For Omeje, first-generation rentier-state authors had limited their scope. By looking only at the state, they had a strong bias for regime stability. They did not see the empowerment of non-state actors. Most important of all, they failed to predict the conflicts which have erupted in these countries. Michael Watts (2008) is one of the first major authors in the literature to use Omeje's rentier space to map out the true causal patterns of conflict in the Niger Delta. This involves building upon the multiple-factor approaches of Mahdavy, Beblawi, and Luciani, by setting beneath these factors multiple contexts.

For conceptual clarity, the rentier space subsumes and upholds activities related to the acquisition and control of oil and

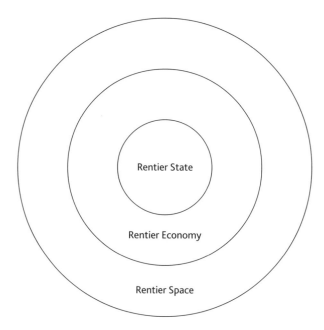

Figure 4.1 Rentier space model

oil-related resources, as well as the disposition, appropriation and utilization of financial resources, dividends and opportunities derived from oil resources. At the epicentre of the coveted space is the rentier state. (Omeje 2006: 6)

Making the state one among many actors, geographically mapping out the playing field, this new conceptualization allows us to describe the potential for non-state actors to pursue resource conflicts, an important problem in Nigeria, Sudan, and Chad, which rentier state theory theorizes only as the long-term political vulnerability of the regime. Both state and non-state actors conflict and cooperate (armed militia, rebellions, secessionists, anti-corruption activists, journalists, lawyers, foreign non-governmental organizations, multinational corporations, organized crime networks) in an "expansive unlevelled playing field underpinned by specific rentier economics, and which subsumes the state and other rentier actors that more or less define the 'rentier space'." (Omeje 2008: 9)

RENTIER MENTALITY, ALLOCATION STATE

For Luciani (Beblawi and Luciani 1987) the key feature of a rentier state is that it is liberated from the need to extract revenues from its own domestic economy. But necessity is the mother of invention. Mahdavy (1970) had observed that the Iranian oil industry's most significant contribution is that it enabled the Shah to embark on large public expenditure programs. Massive spending, without having to resort to domestic taxation or burdensome public debt, should have given Iran a short-cut to development. For Mahdavy the question was why this had not occurred: "Perhaps one of the more crucial problems that needs to be studied is to explain why the oil exporting countries, in spite of the extraordinary resources that are available to them, have not been among the fastest growing countries in the world." (1970: 432–4)

Taking the state's financial autonomy as his point of departure, Luciani classified all states according to their fiscal policies. A "Production State" is one that relies on taxation of the domestic economy for its income. In this kind of state, domestic economic growth is imperative, and so government economic policies are developmental. In an "Allocation State," however, the government does not depend on domestic sources for its revenue, but rather is by itself the primary source of revenue in the domestic economy. Since domestic economic development is not directly related to the government budget, an Allocation State "fails to formulate anything deserving the appellation of economic policy." (Luciani 1987: 70) "Rentier Mentality" is the idea that a rentier economy is premised on and creates a specific kind of mentality. Economic behaviour in a rentier state is distinguished from conventional economic behaviour by a rentier mentality that "embodies a break in the work-reward causation." (Beblawi and Luciani 1987: 52) Rewards of income and wealth do not come to the rentier as the result of work, sacrifice, or investment, but are the result of chance or situation. Mahdavy lamented this fact when he contrasted the somewhat lackluster attitude prevalent in the rentier class with the sense of alarm and urgency prevalent among most other state leaders in underdeveloped countries to alleviate the poverty of their people. "Whereas in most underdeveloped countries, this kind of relative regression will normally lead to public alarm and some kind of political explosion aimed at changing the status quo, in a rentier state, the welfare and prosperity imported from abroad pre-empts some of the urgency for change and rapid growth and coincides with socio-political stagnation and inertia." (Mahdavy 1970: 437) Satisfied with their material conditions, "[i]nstead of attending to the task of expediting the basic socio-economic transformations, they devote the greater part of their resources to jealously guarding the status quo." (ibid.: 443)

"Rentier Mentality" has profound effects on economic productivity. The break in the work–reward causation means that for the rentier "*reward becomes a windfall gain, an isolated fact.*" (Beblawi and Luciani: 52; emphasis added) Income and wealth are seen as situational or accidental, rather than as the end result of a long process of systematic and organized production. Jobs and contracts and licenses are given as an expression of patronage and clientism rather than as a reflection of sound economic rationale. Civil servants see their principal duty as being available in their offices during working hours. Businessmen abandon industrial manufacturing and enter into real estate speculation or other special service sector activities associated with a booming oil economy. The best and brightest seek out lucrative high-paying government posts. Everybody knows getting access to oil-rent is how to get rich. Beblawi concludes that such psychological conditions of the rent-dependency complex represent "*a serious blow to the ethics of work.*" (ibid.: 8; emphasis added)

Another frequently cited problem with oil-dependent economies is that they are highly vulnerable to external price shocks. All oil-rentier states have been affected at one time or another by this Achilles heel, even if their exposure to price fluctuations has been a shared, and not a uniform, experience. Economic diversification varies considerably from one oil economy to the next. But all states in which oil-rent predominates have shared an education in the uncertainty of world oil markets, and have taken measures to protect themselves against future trauma. For example, oil stabilization funds have been set aside using surplus windfall revenues to pay government expenses when prices collapse. Still, looking at the experience of most oil-rentier states, one has a sense that the learning curve has been too short and steep. It is not clear that they have really been able to learn from their own recent past. If the lesson is diversification, that rentier states should become something else, this suggests that the only solution for rent dependency is not to be dependent on rent in the first place. This is not so much a solution to the problem as an admission of it.

How an oil-rentier state can diversify its economy, and into what activity it can diversify, is where the learning curve flattens out. Assuming that it can purchase development with its oil revenues mistakenly assumes that development is a commodity, rather than a process. Examine the chain of causality in a rentier state (Figure 4.1) to see how the inflow of massive oil revenues patterns the problems of economic, social and political development. One major problem observed by Mahdavy was that, "however one looks at them, the oil revenues received by the governments of the oil exporting countries have very little to do with the production processes of their domestic economies" (p. 429). Often the population is too small for local refinement and consumption to make economic sense, but domestic use of oil is limited anyway by the state's rent-seeking export promotion of its crude. Since most oil is produced for export, little is left behind for local refinement or consumption. So petroleum industries in the oil-rentier states tend to be *enclave industries* that generate few backward or forward linkages. Backward linkages are the purchase of local inputs. Forward linkages are the domestic use of output in further productive operations. (Frank 1980: 89) Sometimes the states require progressive increases in the local value-added content through subcontracting to local firms. At other times they decree an indigenization of personnel to increase local participation. But a general lack of inter-industry linkages between the oil sector and the local economy prevents the oil enclaves from becoming launchpads for development.

The mechanism of a rentier economy is premised on the inflow of massive amounts of external rent. This rent comes in a concrete form of foreign exchange (oil is sold for dollars). Access to foreign exchange is important for all developing countries because it allows them to purchase not only consumable goods (food, fuel, medicine, etc.) but also the technology of advanced industrial capitalism (machines, tools, parts) and high-skill services. Many other developing countries must suffer costly balance-of-payment crises or inflation to acquire these goods and services. This should not be the case with a rentier state, an economy saturated in hard foreign currency. But unexpectedly the inflow of external rent on unprecedented scales has time and again tended to throw the input–output matrix of rentier economies into imbalance, as both the state and the society become increasingly dependent on the continual input of this foreign revenue. One consequence is that the state tends to relax constraints on foreign exchange, *imported manufactures replacing domestic manufactures* that lack a sufficient economy of scale. If the rentier state uses its oil revenues to purchase imported foodstuffs, they will also compete with domestic produce on the local markets. Combined with the attraction of rural workers to the urban areas where the oil revenues are concentrated, *oil rents also cause a decline in both agricultural production and rural living standards.*

Another consequence of the availability of large amounts of external rent is that government can embark on big *capital-intensive development projects*. Possessing the foreign exchange required to purchase foreign technology, the rentier state has a capacity to embark on large-scale infrastructural campaigns and state-run industrial complexes. The short-term benefits of such programs are attractive because infrastructural development can employ domestic labour and also because modern industrial complexes endow the state with prestige. The long-term consequences, however, are less impressive. Rather than enlarging the goods-producing capacity of the economy, inter-sector linkages tend to be negligible because of the high import intensity of infrastructural construction activities. The state-owned industries are often worse in that they cannot employ a significant percentage of the population and often demonstrate little commercial viability. They may even drive out small-scale local capital from similar productive activities. These state-owned industries also tend to be enclaves, relying on constant imports for their upkeep and maintenance. They tend to provide expensive advantages for their personnel, such as recreational facilities and round-trip voyages on company aircraft for family visits to keep them from getting too homesick.

So long as oil rent continues to flow from the petroleum sector into state coffers, unprofitable but prestigious development projects may continue to enjoy allocations. Conversely, successful projects may lose government investments when state revenues decrease—unrelated to the success of the projects themselves. Since the oil sector is an industry, deceptive figures showing an absolute increase in "industrial" production may therefore exist parallel with a relative decline in the manufacturing sector. Rentiers can conspicuously consume imported manufactured goods, moreover, without domestic substitution being pursued at all. Spending unearned income, with a rentier mentality, they prefer to purchase imported goods. When they fail to substitute these with locally manufactured goods, this causes a decline in local domestic manufacturing.

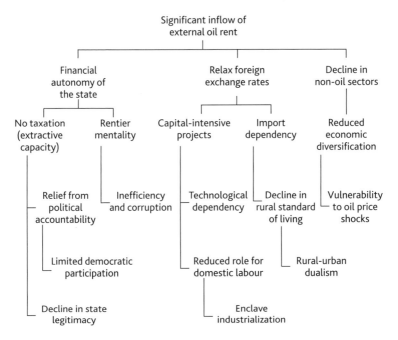

Figure 4.2   Chain of causality in a rentier state

Why does their demand for domestically manufactured goods not keep pace with their demand for imports? There are several explanations. The first is "conspicuous consumption" (Weblen

1899), that is, consumption of goods for purposes of creating invidious comparison. The status conferred by foreign imports often makes them more desirable. The second is that foreign imports often possess inherent qualities resulting from advanced process engineering by foreign manufacturers. The third is that there is a temptation for government to maintain artificially high exchange rates for their national currencies to facilitate the purchasing power of their money. The relative price of imported goods becomes low enough to disadvantage domestic manufacturers not only in the local economy but in the external markets. Export-oriented industry loses its comparative advantage and import-substitution industry loses its economic rationale (real profitability). The fourth is that domestically manufactured goods are often produced in the absence of viable markets. Industries are targeted for development by state policies that have non-market considerations in mind (such as jobs, prestige, symbolism, kickbacks, electoral patronage, or foreign pressures). Prestige-oriented industrialization of this kind is pursued for perceived benefits associated with modernization, rather than a real market demand for the goods. When diversification is pursued for its own sake and the supply of diversified goods is not met with effective demand, domestic industries become net consumers, rather than net producers, of the national income.

The rentier's comparative advantage is his abundance of cheap oil, for which the world markets demonstrate relentless demand. But increased dependency on imports and declines in the non-booming tradable sectors—the "Dutch Disease"—is a pathology that has been observed in the oil-rentier economies afflicted by price shocks. The Dutch Disease afflicts countries with booming oil sectors by distorting the patterns of growth in the agricultural and other tradable productive sectors of the economy (Gelb 1988). It takes its name from the situation in the 1970s when booming North Sea gas exports pumped massive oil rents into the Dutch economy, which appreciated the Dutch guilder and, in so doing, exposed Dutch manufacturers to more intense foreign competition and higher unemployment. In a country with the Dutch Disease, the booming oil sector attracts rural workers away from agricultural production while at the same time contributing to a relative devaluation of local foodstuffs. The same happens in the manufacturing sector. Capital is reallocated to the oil sector, where returns are higher than in either farming or manufacturing. Since the government is the principal recipient of oil rents, there is a tendency for its bureaucracies to expand. Financial services also increase to meet the needs of incoming foreign exchange. Oil service industries also experience distorted growth: pipeline maintenance, storage tanks, port facilities, helicopters, and other transport businesses, and merchants supplying oil company employees in the enclaves with tertiary services and basic supplies (such as restaurants, shops, and hotels).

Map 4.1 Equatorial Guinea

The paradoxical negative impact of sudden oil wealth on an economy can be described as follows: An economy experiencing an export boom can be divided into three sectors, which are: (1) the booming export sector, i.e. oil; (2) the lagging export sector of traditional exports; and (3) the non-traded goods sector. In the presence of Dutch Disease, the traditional export sector gets crowded out by the other two sectors. Oil windfalls lead to an appreciation of the real exchange rate by shifting production inputs (capital and labor) to the booming mineral sector and non-tradable sector (retail trade, services, and construction), thereby reducing the competitiveness of the non-booming sectors of agriculture and manufacturing, hence causing their collapse. Consequently, rural standards of living decline. The Dutch Disease is a matter of one sector benefiting at the expense of others.

SPANISH GUINEA (1778–1968)

Equatorial Guinea is composed of two very different parts: a mainland territory known as Rio Muni, and several islands, the most important being Bioko, with its capital city Malabo. This large island, crossed by two chains of volcanic mountains, has abundant water supplies, fertile volcanic soil, and three natural harbours. But the Portuguese who "discovered" these islands only used them as a slave entrepôt, and to resupply their ships with fresh food and water. They eventually ceded them to Spain in the Treaty of Pardo (1778) and the Spanish took Rio Muni in a joint expedition with the French in 1900. Over the nineteenth century, as Spain declined in power, it never managed to establish a settler colony. There were several attempts to settle with colonists. All of them failed. One attempt in 1859 failed so spectacularly that six months after the colonists had left Spain only three individuals could be found still living on the island. All the rest had either perished or had returned to Spain. The main cause of death was malaria, called "African fever," for which they had no cure. For three centuries the islands had no permanent Spanish settlers. The only whites were colonial officers and timber merchants, who usually served a term of a few years; then returned home. In 1845 the British had actually offered to buy the islands, but when the King of Spain presented this plan to the Cortes for their approval, it was soundly rejected by the legislators. For Spain's only African possession, even if they had not truly colonized it, had by then become a very important symbol of national pride. After the Berlin Conference (1884–85) "Spanish Guinea" was a negligible crop of islands, with 28,000 square miles of largely unexplored land onshore around the River Muni.

In terms of natural endowment however it was a tropical paradise. We have an eyewitness portrait in the mid-nineteenth century painted by the French Consul: "A magnificent panorama unfolded before any voyager's eyes who arrived by sea," he wrote. "Above them rises Isabelle Peak, at a height of 2886 meters above sea level, invaded by a thick forest at the base of the summit. Around the mountain's feet arises the city of Sainte-Isabelle, a few hundred meters above sea level, also enveloped by ageless trees." (Benedetti 1869: 71) He extolled its many private gardens, "where the orange and palm trees are always blossoming," (ibid.: 72) and the generally good conditions for agriculture. "The soil, remarkably fertile, is capable of furnishing products from the tropical zone, as well as the temperate zone and cold countries, according to the region of the island one is considering." He praised the island's many waterfalls: "Water is abundant and better than one finds on the West coast of the African continent." (ibid.) Most of all he praised its suitability for cash crops. "All of the tropical products succeed: tobacco, coffee, cocoa, cotton, sugar. The most common fruits are the orange, the lemon, the guava, diverse species of banana, pineapples, and coconuts; all introduced onto the island since 1827, some by the English, others by the Spanish." (ibid.: 72–3) The palm tree was the most important species of tree for agricultural production, providing the palm oil traded by the indigenous people, the "Bubi."

The Bubi were estimated in those days to number somewhere between 4,000 and 20,000 people, and their unique commercial occupation consisted of extracting palm oil. In general they lived in the middle of the forest in cabins grouping together 40 or 50 families. "They are completely different from the inhabitants on the neighbouring continent," observed Benedetti. "They are more copper-coloured than black. They are small in size, odd-shaped. Men and women walk around with no clothes." (ibid: 69) Despite their strong work ethic, "the quantity of palm oil extracted could be more abundant," complained Benedetti, "without the indolence of the natives. [...] No other export but palm oil, around 200 to 300 tons, comes out of the port of St.-Isabelle each year." (ibid.: 77)

The natives of the other islands—Annobon, Corisco, and Elobey—traded eraser gum, stained wood, and ebony with the peoples living along the neighbouring coastline. The islanders brought goods in canoes across the Gulf of Guinea and went upriver trading these goods. By the mid-nineteenth century Annobon was inhabited by a few thousand ex-slaves who spoke an African dialect of Portuguese similar to Sãotoméan and Angolar. Their island was so poor in resources that they had nothing to trade with Europeans except items from the mainland. Annobon was sterile, and lacked safe anchoring, so it was rarely visited by the merchant marine, and even when they captured whales, these had to be processed in Gabon. Corisco, near present-day Gabon, had a population of a thousand Bengas, and a handful of Jesuit missionaries from Spain. It had served as a slave entrepôt before abolition, but its sandy soil was equally unsuitable for farming. Of the four islands, Fernando Poo (later renamed Bioko) was by far the most important. It was the only island where any Spaniards really ever settled, so the "Fernandino" Bubi were more assimilated.

However, the territory on the continent, Rio Muni, was the largest part of the colony, and contained most of the population. Here the native peoples were the "Fang." For the first one hundred years of Spanish colonial rule, they governed themselves. The Fang were Bantu people who had migrated forcibly into the territories of present-day Cameroon, Equatorial Guinea, Gabon, and Congo; they were organized politically into clans that had pushed into the rainforests. Legend says the pygmies taught them how to hunt elephants. The Fang sold ivory tusks to coastal middlemen who in turn traded with the Europeans, until they decided to invade the coastal regions to trade directly with the Europeans themselves. The immense army of clans has been described by the economist Oyono Sa Abegue (1985) as a kind of "military democracy," with warlords simultaneously assuming magico-religious and judicial functions. At the summit of their political order was a supreme military chief, supported by a council of elders and a people's assembly. This ruler was supported so long as he redistributed the wealth among his clansmen successfully; he could be removed from power by the council of elders and the people's assembly. Although hereditary lineages did emerge, the selection process tended to favour individuals who possessed the most prowess.

By the end of the nineteenth century, when Spain began to effectively colonize its last African possession, the traditional patrimonial systems were replaced by a colonial tyranny. At the summit of their new political order was the Governor General, a Spanish officer given the *ex-officio* rank of brigadier general. His was a modern legal–rational authority system, with a bureaucracy, but no democratic accountability. Laws, tribunals, and prisons were conceived to exclude the indigenous populations from any posts of responsibility. The Spanish colonial experience is a pertinent base for understanding contemporary Equatorial Guinean institutions, for it established a model of governance that resembled Spanish fascism. The Africans who came to power at independence were soldiers, trained by Spanish fascists. They did not fight an anti-colonial war of independence, but rather negotiated decolonization with the Franco regime, and established an authoritarian system of government that strongly resembled their fascist masters. Moreover, because the island capital had been the focus of almost all Spanish assimilation, the Bubi had enjoyed a relatively privileged status during the late colonial period. This changed after decolonization. The formerly underprivileged Fang, by virtue of their numerical preponderance, assumed political power, and thereafter tyrannized the smaller ethnic minorities who inhabited the islands. Later, when oil was discovered offshore, their fascist-inspired system of government metastasized into a repressive and tyrannical rentier state.

DICTATORIAL GUINEA

During the eleven years following independence of Equatorial Guinea in 1968, the country was dominated by the dictatorship of Francisco Macias Nguema and his Mongomo Clan of the Fang, who devastated the economy and brutalized the population. The economy inherited from the Spanish colonizers was based almost exclusively on a primary sector comprising agriculture, fishing, and forestry, which accounted for 50 percent of GDP, 97 percent of exports, and 80 percent of the income for the population. (World Bank 1983; cf. Same 2008: 4) In addition to its infamous cruelty and madness, the new regime was characterized by weak public management. The economy was mismanaged, most public administration ceased to function, and the population survived at a barely subsistence level.

Francisco Macias Nguema was a paranoid schizophrenic sociopath who declared himself president for life. The cinematic depravity of his regime was legendary. In 1975, for example, he celebrated Christmas Day by lining up 150 of his political opponents in a soccer stadium and shooting them dead while a macabre brass band played "Those Were the Days My Friend." (Shaxson 2008: 34) "On another occasion, thirty five prisoners were told to dig a ditch and stand in it. The trench was then filled so that only the men's heads stood out of the ground. Within twenty-four hours, ants had slowly eaten the prisoners' heads, and only two men remained alive." (Ghazvinian 2007: 17–2) Most of the tiny educated class was killed, approximately one-third of the population fled the country, and the formal education system ceased to function. As a result of the madness of Macias' regime, GDP per capita fell from $260 in 1970 to around $170 in 1979. Following the departure of foreign plantation owners, cocoa exports fell from nearly 40,000 tons in 1968 to less than 20,000 tons at the beginning of the 1970s, and then to about 7,000 tons by the end of the regime. (World Bank 1984; cf. Same: 4) In an attempt to restore cocoa production, the government nationalized most plantations, introduced forced labor, and brought large numbers of people from the countryside to work on the state-owned plantations. These workers were of course unmotivated, the management of their activity was poor, and the reduced application of insecticides and fungicide led to a further decline in yields in the remaining area under cultivation. Exports fell to a record low of 5,200 tons in 1980. (ibid.: 4–5) When the government started seizing harvests without payment, cash crops were completely abandoned and barter trade became the dominant form of exchange.

Timber production experienced an equally stark decline, from 360,000 cubic meters in 1968 to an annual average of 6,000 cubic meters in the late 1970s. Meanwhile coffee and palm-oil production virtually disappeared. Basic services in health, education, water, and electrical supplies could not be maintained, foreign investment stopped, and the trading system, operated by state enterprises, broke down. (World Bank 1986; cf. Same: 5) Devastation of the economy in the 1970s was accompanied by complete disarray of public finances. "Public financial transactions were recorded only sporadically, and the accounts of the Treasury, the Bank of Equatorial Guinea— the former Central Bank—and public enterprises were not kept separately." (ibid.) Macias Nguema was finally overthrown in a 1979 coup d'état led by his nephew Teodoro Obiang Nguema, the military governor of the island and director of its infamous Playa Negra prison. It is said that Macias fled into the forest with a suitcase full of cash containing the entire national treasury. Surrounded in a cabin hideaway and unable to escape, he reportedly burned this money in a final act of mad vengeance (Klitgaard 1990), before being captured, tried, and executed by his nephew.

Since 1979, Obiang Nguema and his military junta have run one of the most despotic tyrannies on the African continent. Although it seems hard to believe that anything could have been worse than his uncle Macias' reign of terror, keep in mind that Obiang had been the head of state security under the old regime, and continued its bloody policies once he took power himself. (Liniger-Goumaz 1989, 1997, 2000, 2005) What makes this hard to accept, especially for those who focus narrowly on the macro-economic figures, is that the country has enjoyed one of the highest rates of economic growth on the continent. Ever since oil started flowing, this tiny country of around one half-million inhabitants has been reporting rising per capita income figures. Those who are able to read between the lines of such fictional averages (which ignore the unequal distribution of oil wealth) are nevertheless likely to have read glowing reports about the regime in glossy special issues of *Jeune Afrique,* orchestrated by public relations firms and paid for by oil revenues. But a glance at human rights reports will quickly reveal that the arrival of big oil has enriched its kleptocratic rulers, funded the oppression of a miserably impoverished people, and maintained a brutal police state behind a façade of slick public relations paid for by oil.

The oil itself is located offshore, and would have belonged to the Bubi had they been granted a separate independence from the Fang. In 1995 ExxonMobil and Ocean Energy discovered the "Zafiro" oilfield, which is located northwest of Bioko Island. Zafiro was the first deepwater field to be brought on stream in West Africa and is currently the main producing field in Equatorial Guinea. It is operated by an ExxonMobil-led consortium that includes Devon Louisiana and the national oil company GEPetrol. Its reserves are estimated around 400 million barrels, and with an output of 245,000 barrels per day. "Ceiba," located just offshore of Rio Muni, is the second major producing oilfield. It is estimated to contain 113 million barrels of reserves and began production in 2000. The field is operated by Amerada Hess, with partners Tullow Oil and GEPetrol. "Alba," the third largest field, is located twelve miles north of Bioko. It is a major condensate field, that is, located in predominantly gas-bearing reservoir rocks, and contains an estimated 400 million barrels of liquids. It currently produces between 65,000 and 75,000 barrels a day of condensed gas and 20,000 barrels per day of liquefied petroleum gas. Marathon Oil serves as the operator of Alba field along with its partner GEPetrol. (EIA 2007) In addition to oil, Equatorial Guinea has 1.3 trillion cubic feet of proven natural gas reserves. The majority of these reserves are located offshore of Bioko, primarily in the Alba and Zafiro associated natural gasfields. Alba contains 1.3 trillion cubic feet of proven reserves, with probable reserves estimated at 4.4 trillion or more. Marathon Oil is the operator and has a 63 percent interest in the field, while Noble Energy holds 34 percent and GEPetrol 3 percent. Natural gas production is around 46 billion cubic feet per year, and the country is currently marketing itself as a regional gas industry hub based on the recent completion of a liquefied natural gas facility on Bioko. (EIA 2007)

To collect its share of the oil revenues the government created a national oil company that became operational in 2002. GEPetrol's primary focus is to manage the interest stakes of the government in its various production-sharing contracts and joint ventures with foreign oil companies. This national oil company can also participate in foreign exploration and production activities.

Following a 2005 decree by President Obiang, the regime also created a natural gas company, Sociedad Nacional de Gas de Guinea Ecuatorial (Sonagas), whose responsibilities include managing gas assets and developing an industrial–residential gas market, as well as treatment, distribution, marketing, and export. (EIA 2007) But they are really just rent-collectors—operational control of the high-tech offshore industry is entirely in the hands of foreign oil companies. Without the help of these foreign firms, neither GEPetrol nor Sonagas would be able to exploit any of their offshore reserve, nor have any technical ability to measure or control the actual amounts of oil being produced.

Is Equatorial Guinea a "rentier state?" Unfortunately, the government provides the public with no information about its current budget and financial activities during the course of the budget year, and has the lowest possible ranking on the Open Budget Index of 0 percent. (Open Budget Initiative 2009) Obiang himself has gone on the record to state that Equatorial Guinea's oil resources are a "state secret." (Global Witness 2004: 61) So we only have estimates of the percentage of government revenues generated by the oil sector. These are from reputable sources. According to the Energy Information Agency, for example, oil revenues increased one-thousand fold, from $3 million in 1993, to $3.3 billion in 2006. According to the IMF, hydrocarbons account for about 90 percent of GDP, 98 percent of exports and over 90 percent of government revenues. (Same 2008: 9) Therefore it is fair to call this a rentier state, since no other sources of government revenue even come close to those provided by oil and gas.

Is Equatorial Guinea ruled by a "rentier class?" It is ruled by the Mongomo clan of the Fang. In all, 21 of the country's 50 cabinet members are direct relatives of the president, with one son, two sons-in-law, one sister-in-law, three brothers, three brothers-in-law, one mother-in-law, two uncles, two cousins, and one nephew holding government accounts at the former Rigg's Bank in Washington D.C. (Liniger-Goumaz 2005: 206) Another son by a marriage with a São Toméan, Gabriel Nguema Lima, serves as the vice-minister of mines, industry, and energy. One of Obiang's brothers, Armengol Ondo Nguema, runs internal security. This is all evidence that the Mongomo Clan of the Esangui Fang (who represent only 1 percent of the population) are a highly concentrated rentier class.

Although the oil and gas reserves are located offshore around the archipelago, the Bubi and other island peoples have had their resources brutally expropriated by the mainland Fang. The Bubi, it should be said, have demanded an autonomous status since before independence, refused first by Madrid, then later by the Nguemists. When they formed a political party, Movimiento para la Autonomía de la Isla de Bioko (MAIB), in 1993 the regime refused to recognize it. When they tried to assault military bases in 1998, the Fang junta arrested 550 Bubi activists, and massacred 150 innocent civilians in their villages. Soldiers patrolled the streets of Malabo, indiscriminately beating and raping their women. "Some of the women had forks thrust in their vaginas and were told, "From now on, that's your husband." (Global Witness 2004: 66) After the initial deaths of numerous Bubi prisoners, without autopsy or investigation, 110 remained incarcerated at the infamous *Playa Negra* prison, where, according to one human rights report, "a large number were submitted to interminable tortures, attested by the wounds all over their bodies, arms and legs." (Liniger-Goumaz 2003: 179) According to the 1999 U.S. State Department Human Rights Report, "Police urinated on prisoners, kicked them in the ribs, sliced their ears with knives, and smeared oil over their naked bodies in order to attract stinging ants," all of this directed personally by Obiang's brother Armengol, "who taunted prisoners by describing the suffering that they were about to endure."

Does the government "lack political accountability?" It is important to remember that Obiang came to power by violence, and has never been held accountable for his coup d'état. He staged a plebiscite in 1982 making himself president and introducing presidential immunity for all his acts. He then inaugurated a new unicameral Chamber of People's Representatives in 1983, whose members he personally appointed. In 1987 he created a one-party regime under the Partido Democratico de Guinea Ecuatorial (PDGE), which he also headed himself. Only members of the PDGE were permitted to hold seats in the legislature. There is an absence of any basic distinction between executive, legislative, and judicial powers. The judiciary is not independent of the executive; judge and magistrates are all named by the president, and few have any formal education in law. Defense attorneys who dare have no access to their clients in prison, who are tortured with impunity. Extrajudicial killings are routine. Obiang created a High Council of the Judiciary (presided over by himself) to oversee the work of all judicial officers. Freedom of speech, the press, and assembly are not protected. A 2003 country report by the International Bar Association concluded that the laws are not written, never properly or consistently used, or are inconsistent with the constitution, outdated, or *ad hoc*. (cf. Global Witness 2004: 66–7)

Is "democratic participation" limited? The U.S. State Department dryly observed in 1987 that "citizens have only a hypothetical right to change their government by democratic means" and twelve years later in 1999 admitted that "in practice, there have been no free, fair and transparent elections." (Liniger-Goumaz 2000: 248, 236) In Obiang's first plebiscite in 1982, voting was obligatory, the army manned the polls, the red ballots said "yes," and the black ballots "no." The vote was not secret. People did not even know the contents of the referendum for which they were voting. Terrorized, 139,744 voted yes, and 6,149 voted no. This led the way to a single-party regime under the PDGE, a situation that started to change, if only superficially, under U.S. government pressure. Obiang staged a second plebiscite in 1991 (98 percent voted "yes"), which granted the right of opposition parties to exist, and eliminated the obligation to vote. Multiparty legislative elections were held (abstention 80 percent). The PDGE took 62 out of 80 deputies. The 1996 presidential election, in which Obiang declared himself the winner with 98 percent of the vote, was considered openly fraudulent by international observers. "Some opposition politicians who campaigned were beaten and jailed. Voting was done in the open and without secrecy, with opposition parties allegedly being barred from access to polling areas. There were credible reports of widespread arrests and violence against opposition party members before the elections, as well as of beatings, roadblocks, stuffed ballot boxes, and the presence of security forces." (ibid.: 237) The 2002 presidential elections were also fraudulent. All but one of the opposition candidates boycotted the elections, and Obiang took 97.1 percent of the vote. In 2009 Obiang declared his re-election with 95.4 percent.

Does the regime suffer from a "rentier mentality?" In a rentier state reward is given not for talent or hard work, but for one's position and connections. The government created an Agency for the Promotion of Employment, which retains two-thirds of employees' salaries as a fee, screens out potential employees considered "unfriendly" to the ruling party, and fires those who complain about ill treatment. Those who want to work for the oil companies have to go through employment agencies run by Obiang's brothers, Armengol and Mba Nguema, Multi Service System (MSS) and Servicio Nacional de Vigilancia (SENAVI). (*El Mundo* 2003) The *Los Angeles Times* revealed that $500 million of the oil revenues were deposited into a bank account of the Rigg's Bank in Washington D.C. exclusively under Obiang's personal control. (Silverstein 2003) Obiang's first son Teodorin is known to have purchased a $6 million mansion in Bel Air, and another $35 million mansion in Los Angeles, from which he runs a hip-hop music label, TNO Records. (Global Witness 2009) Oil companies have also taken advantage of the regime's preference for immediate cash by loaning money for what the IMF obliquely calls "unrecorded extra-budget spending" at non-concessional rates (IMF 2003: 3), providing up-front loans at high interest and then deducting this money at the source from oil payments. (Global Witness 2004: 69) This is evidence of the preoccupation of a rentier class with its own immediate consumption rather than long-term development.

Has the regime experienced the "Dutch Disease?" First it must be noted that the country abandoned its national currency for the CFA franc. Nevertheless, the real effective exchange rate of the CFA franc calculated for Equatorial Guinea has continued to appreciate since the coming of oil revenues. This pattern largely reflected the appreciation *vis-à-vis* the US dollar and the persistent large inflation differential with the major trading partner countries. According to a recent policy research working paper published by the World Bank there is ample evidence of Dutch Disease: "The appreciation of the real exchange rate for Equatorial Guinea was the largest in the CEMAC region, reflecting the country's position as the largest oil exporter and recipient of foreign direct investment relative to GDP." (Same 2008: 10) Another crucial underlying factor for stagnation in traditional exports was the loss of immigrant farm labor, "which was exacerbated by the abandonment of the farms in search of more lucrative employment in the oil and related sectors." (ibid.: 10–11) The report concluded, however, that the negative impact of Dutch Disease was limited given the structure of the economy, and on the contrary may even be a good thing because it fuels the structural transformational process of the economy. It argued that the ongoing Dutch Disease is a "natural and necessary" reallocation of resources: "In a country where the manufacturing sector barely exists or where the non-oil primary sector is structurally deficient, there is little to fear about the disease. The oil boom is a blessing, given that oil revenues when properly managed can play a special and critical role in overall economic development and poverty reduction." (Same 2008: 1)

What is most disturbing about this rich, detailed, empirical report (providing all of the key economic data over the entire period of the oil boom until now) is how blind it is to the political realities of the country. There is not even the slightest bit of evidence that oil revenues are being "properly managed." Of course, proving that Equatorial Guinea suffers from the syndrome of the rentier state is a little like proving that someone with leprosy suffers from a skin rash. Billions of dollars of oil rent did not make this country worse. But they did not make it better, either. Many people falsely believe that if you pour enough money on an African country it will eventually develop its economy and even become more democratic. The reason that the theory of the rentier state is important to know about is that is shows how massive oil revenues, even an oil boom, do not generate meaningful development, and moreover, can prevent such a regime from embarking on positive political change. Oil money has funded a bloodthirsty dictatorship, and *that* is its major contribution to Equatorial Guinea.

5
Praetorian Regimes and Terror

SOLDIERS AND OIL

Almost all African oil states have experienced coups d'état and military rule (Table 5.1). Of course, on statistical grounds alone, coups d'état and military regimes are the two most prevalent political phenomena in Africa. But if soldiers must always publicize their coups in order to assume power, they do not like to draw attention to the military nature of their regimes, for the military regime is no longer considered a legitimate form of government. So the soldier always promises that his rule is transitional. He must promise he will hold elections, or at least restore civilian order. When he does so, the international community applauds. When he does not, sanctions are imposed, aid is cut, and his junta must eventually be "civilianized." That is, soldiers camouflage themselves with civilian titles, move out of their barracks, stage elections, occupy ministries and give an appearance of civilian rule. Sometimes, however, we will find officers who prefer to wear their traditional military uniform, to receive their promotions in rank, and to run their country from the army barracks. But all those who remain in power long enough have eventually moved into the presidential palace and exchanged their khakis for the costume of a civilian "president."

Table 5.1 Coups d'état and civil wars in African oil states

	Coups d'Etat	Civil Wars
Angola		1975–2002
Cameroon		
Chad	1975, 1990	1960–90
Congo-Brazzaville	1963, 1968, 1977, 1979	1993–97
Equatorial Guinea	1979	
Gabon	1964	
Mauritania	1978, 1980, 1984	
Nigeria	1966, 1975, 1983, 1985, 1993	1967–70
São Tomé & Príncipe	1995	
Sudan	1958, 1964, 1969, 1985, 1989	1956–72, 1982–

Box 5.1   Why military regimes "civilianize"

"Politically the armed forces suffer from two crippling weaknesses. These preclude them, save in exceptional cases and for brief periods of time, from ruling without civilian collaboration and openly in their own name. Soldiers must either rule through civilian cabinets or else pretend to be something other than they are. One weakness is the armed forces' technical inability to administer any but the most primitive community. The second is their lack of legitimacy: that is to say, their lack of a moral title to rule." (p. 14)

S.E. Finer, *The Man on Horseback: The Role of the Military in Politics* (1962)

Three different schools of thought explain the causes of military coups. The first is they are caused by *structural* weaknesses such as institutional fragility and/or "minimal political culture." (Finer 1962). Typical of this approach was the early work of Samuel Huntington (1968), who argued that "the most important causes of military intervention in politics are not military but political, and reflect not the social organizational characteristics of the military establishment but the political and institutional structure of society." (p. 194) Soldiers rule over weak states. The second school teaches us that the military possesses certain *organizational* characteristics of professionalism, nationalism, cohesion, and austerity that lead them to move into the political arena and rescue the state from corrupt self-seeking politicians. (Janowitz 1964) Their army training molds an officer corps into cohesive, non-tribal, disciplined, and national units. Since the military is the most modern, Westernized, and efficient organization in society, it rules over less modern parts of society. The third school of thought places more weight on the *personal* motives of ambitious or discontented officers. In one of the most famous studies of this school of thought Samuel Decalo (1976) argued: "It is both simplistic and empirically erroneous to relegate coups in Africa to the status of a dependent variable, a function of the political weakness and structural fragility of African states and the failings of African civilian elites." (p. 13)

Needless to say, links between structural characteristics of political systems and the incidence of military takeover will not appear if one of the key variables—the *idiosyncratic* element, or personal ambitions of military officers—is not taken into account. Moreover, knowledge of the true motives of military officers in overthrowing civilian regimes can give us insights into the kinds of policies they are likely to follow once in office, while misjudgments of their motivations will lead us to uncalled for expectations of military rule. (Decalo 1976: 22)

It is not the purpose of this chapter to settle a theoretical debate over the causes of military coups. Nor does it matter whether a military ruler came to power by a sudden coup d'état or over a *long* civil war. What matters are the effects of the relationship between oil and military rule, if only because so many of the African oil-exporting countries are ruled, overtly or covertly, by professional soldiers. Six out of eight (or 75 percent) of the rulers in African oil-rentier states are soldiers by profession, and came to power by coup d'état or by winning a violent civil war. This is much higher than the overall average for Africa, where 22 out of 52 rulers (or 42 percent) came to power by coup d'état (Libya's Gaddafi, Equatorial Guinea's Nguema, Guinea's Konate, Burkina Faso's Compaoré, Chad's Déby, Gambia's Jammeh, Uganda's Musevini, Guinea-Bissau's Sanha, Madagascar's Rajoelina, CAR's Bozizé, Mauritania's Aziz, Sudan's Al Bashir); through violent civil war (Angola's Dos Santos, Zimbabwe's Mugabe, Eritrea's Afwerki, Congo's Sassou-Nguesso, Rwanda's Kagame, Somalia's Ahmed, Ethiopia's Zenawi) were installed by the military (Togo's Gnassingbe, DROC's Kabila) or by secret police (Djibouti's Guelleh).

Oil was neither a necessary nor a sufficient condition of military rule in Africa. It was not, strictly speaking, the *cause* of these military dictatorships. There are military rulers in countries that don't have oil, and there are oil-dependent countries not ruled by the military. Also, the same country had alternating periods of civilian and military rule. As one scholar noted over 40 years ago, when statistically correlating the structural characteristics of regimes that had suffered military coups, "it is impossible to specify as a class countries where coups have occurred from others which have so far been spared." (Zolberg 1966: 71) This led Decalo to conclude that the search for the structural causes of coups was erroneous. "The core analytic flaw is the confusion of very real and existing systemic tensions in African states (which are, however, the universal *backdrop* of all political life on the continent) with other factors— often the *prime* reasons for a military upheaval—lodged in the internal dynamics of the officer corps." (1976: 13) Oil dependency is a contextual factor where military rule is empirically more probable in Africa (75 percent to 42 percent), not a cause *sine qua non*. So if you were to remove oil, you would not necessarily end military rule. Nor is it a sufficient cause. The mere presence of oil is insufficient to produce a coup d'état.

So the purpose of this chapter is to examine the *effects* of military rule on oil-dependent countries in Africa, particularly the possibilities of such a regime to use its oil revenues to alleviate poverty or promote development. This book is about how to stop Africa's oil curse. Its problem is not military rule, *per se*, but oil. The question is whether soldiers are capable of defending their resources from foreign powers (Chapter 1), resisting oil multinationals (Chapter 2), adopting international good governance initiatives (Chapter 3), and/or combating the negative effects of oil rents (Chapter 4). If they are, then perhaps we ought to help these regimes to help their people. If not, then we must look for a better solution below the state level of analysis. The following sections will first describe in theoretical terms a typology of different kinds of military rule, and provide the predictable consequences of each type of military regime on the political and economic development of their countries. Then the latter sections will test this typology and its theoretical propositions with a critical case study of oil-funded civilianized military rule under Denis Sassou-Nguesso of Congo-Brazzaville.

TYPOLOGY OF MILITARY STYLES

It is a commonplace assumption that all military regimes are alike. This follows from common sense. Soldiers are the ones who stage the coup d'état. Usually an officer takes power, often still wearing the outwardly visible signs of the military uniform. He is surrounded by armed subordinates whose insignia indicate rank and hierarchy, who travel around the country in military vehicles, who employ military tactics to establish political order, and who sometimes rule directly from the barracks. All military regimes start this way, with a certain set of shared characteristics that clearly identify them as being "not-civilian," which is why we qualify them as "military" regimes in the first place. But to presume that all military regimes are alike because they are not-civilian would be the same as presuming that all men are alike because they are not-women. Military regimes are as different as the men who create them and the societies over whom they rule. They are as different as the historical periods in which they come to power, and the structure of the international system in which they must survive. Therefore it is important to create a typology that allows us to distinguish one kind of military regime from another, so that we can predict the different developmental outcomes of various "styles" of military rule.

First we must disregard the outward signs of military rule. When a military dictator takes off his uniform and puts on a presidential suit and tie, he is still a military ruler. His change of costume is simply a kind of camouflage that lets him operate in civilian surroundings. The "civilianization" of military rule is a timeless disguise, and goes to the very heart of what it means to be a soldier. What is a soldier? Before the professionalization of modern armies, a soldier was anybody who found himself in an army or navy. Even today, many civilians find themselves conscripted into military service for a limited duration of time. While the ancient historians may have described armies as social structures, military service has usually been a temporary function. It is a type of service, not a status. To identify someone as a "soldier" is to designate what he does (his function), and not who he is (his being). Of course, there has always been a type of person that we could call a "warrior." Even ancient mythology testifies to the eternal existence of people whose specialized function in society was to kill or be killed. It is further obvious that history has properly designated individuals as being part of larger social groups called "armies" or "navies." But whenever we study them biographically, we find that individual soldiers are people who come out of society, that is, they are civilians serving a military function.

Therefore if we want to create a modern typology of military regimes, we should avoid the idea that a soldier is a kind of thing, and then deduce his motives from that reified abstraction. Rather we must look at the soldier concretely as human being. Then we will understand his true motivations, the real sense of his actions, and the intended consequences of those actions. By separating the individual level of analysis from the organizational level, we can also assign moral responsibility to violent acts committed under military rule. Some acts committed by a soldier are determined by his chain of command, that is, by the principle of hierarchy inherent to any military organization. In such circumstances the old principle of *respondeat superior* should be applied. The one giving the command is morally responsible: "He who acts through another, acts for himself." There are other acts of a soldier, however, which are not determined by his organization, but are based on his own free will. Here the moral responsibility of a soldier is the same as that of a civilian: He is responsible for all the intended consequences of his volitional acts. To hold an individual morally responsible for the acts of his organization, or to hold an entire organization responsible for acts of one of its members, is to commit a logical fallacy of amalgam.

We are interested in theorizing a typology of military regimes in order to generalize about predictable outcomes within an oil rent-dependent context. We are not, however, interested in creating a psychological typology of individual soldiers. This is a work of political science, not political biography. Therefore we must classify the various types of military organization. This is where the idea of "military style" is particularly useful. Military "style" refers to a modality of rule, i.e. the principal systemically relevant features of military behavior in office. It is an idea first enumerated by Samuel Decalo (1976), who criticized those apologists and advocates of military rule who imagined African armies as modern complex organizations resembling the military in their own countries. Most African armies bear little resemblance to a modern complex organization model, but are instead "a coterie of distinct armed camps owing primary clientist allegiance to a handful of mutually competitive officers of different ranks seething with a variety of corporate, ethnic, and personal grievances." (Decalo 1976: 14–15) The "personal" element explained more than any of the "structural"

Box 5.2 Six dimensions of military "style"

"To serve its purpose a typology should be capable of differentiating between political systems that share secondary characteristics. There are a large number of components that make up an elite's particular style of rule, and diverse modalities of military rule may be constructed along several of these dimensions. [...] Of the various characteristics of military rule, the following six may be suggested as having prime importance: the corporate status of the armed forces, the permeability of civil–military boundaries, the degree of personalist concentration of decision-making authority and coercive power, the satisfaction or non-satisfaction of nonmilitary group demands, the relative immunity of the regime from personality cleavages leading to praetorian assaults, and the active-combative or passive-reconciliationist approach of the regime to society issues." (Decalo 1976: 241–2) explanations of coups, such as the official ideology, the level of development, the level of ethnic fragmentation, or even the strength of institutions. Scholarly literature of his day was saturated with neat two-dimensional typologies that failed when applied to the empirical diversity of military regimes. Apart from the fact that such generalizations did not hold true when they were superimposed on specific armed hierarchies in Africa, he found that a random *idiosyncratic* variable was always operative, and called this the "personal style" of the military ruler.

Using six abstract dimensions—or characteristics—of military rule, Decalo distinguished five different styles. The first he called "Holding Operations." These are military regimes whose credentials are transitory. They are temporary administrations, so they tend to be stable. The boundaries between the military and the civilian functions are not overtly destroyed by intrusion into the political arena. Soldiers remain soldiers, and do not become politicians or civil servants. They remain in the barracks, so to speak. Thus the corporate status of the military as an institution within society remains relatively high, in part because "[s]elf-aggrandizement by officers in political office tends to be of modest proportions, and the corporate unity of the army is not usually threatened." (ibid.: 248)

The second kind of military style he called "Managerial Brokerage." The distinctive feature of this kind of regime is the attention it pays to competing group demands within society. It is a kind of arbitrator between civilian factions who are competing for power and resources. This kind of military regime is the ultimate broker in the decision-making process. Its role is to ensure that no group (including the army itself) gains a stranglehold over all the patronage, largesse, or state resources. It tends to be relatively stable in that it actively caters to the satisfaction of group demands outside the armed forces, and therefore "is able to build a modicum of popular support or consensus." (ibid.: 247)

The third kind he called "Bureaucratic," a name borrowed from scholarly literature coming out of Latin America. It exhibits less efficiency or success than managerial brokerages, and it is unclear about the transitory or temporary nature of its power. It arises in countries where the status of the armed forces is not very high (making brokerage difficult) and the boundaries between civilian and military functions are quasi-permeable. Soldiers become bureaucrats and take over civilian ministries and corporations. The concentration of power is low at the center, so the army is rule-bound and bureaucratic, not run merely through personal loyalties to its leader. Bureaucratic style can be considered a less successful form of Managerial Brokerage in which the "satisfaction of group demands is less evident than in brokerage systems." (ibid.: 250)

The fourth kind he called a "Personal Dictatorship." In this type of rule "power becomes centralized in the hands of one leader, who uses selective terror and purges to maintain himself in office." (ibid.: 24) A military dictator rules, more or less, in an imperial style. He makes himself emperor, with all the benefits of patronage, largesse, and authority flowing directly from his throne. The Personal Dictatorship is the ultimate in political decay, since such rule often involves terror and bloodshed, and the leaders fear retribution if they lose power. A return to civilian rule or the creation of political institutions is highly unlikely in this kind of military regime.

The fifth kind he called "Praetorian." This kind of system is typified by intense intra- and inter-elite strife, "the presence of continuous plotting and jockeying for supremacy within the ruling junta, and a permanent tug-of-war for influence and power between various groups and military factions." (ibid.: 243) The distinguishing feature of Praetorian military rule is its acute instability, coupled with a high level of permeability of civil–military boundaries. Here soldiers siege, occupy, and pillage civilian administrations, transforming them into semi-feudal autonomous fiefdoms. Of all the types of military rule, this kind of system tends to produce the very lowest level of achievements in both the socio-economic *and* political arenas.

The main purpose of his study was not to evaluate the merits of military vs. civilian rule. Nor was Decalo advocating some perverse kind of "good military governance." He wanted to underscore the weakness in a number of theoretical constructs about African armies by applying them to empirical realities. Objective reality was much more complex than neat two-dimensional theories. It was the manner by which a military regime ruled, and not the mere existence of a military regime, that he believed to be the proper object of scholarship. Since his six variables are all quite explicit in terms of the characteristics of military rule they defined, they "should not offer any insurmountable difficulties to the empirical researcher wishing to operationalize them in order to set up preliminary ordinal scales against which all military regimes may be scored." (ibid.: 242)

In other words, after conducting empirical research and classifying a regime into one of the five types, it is then possible to make predictions about the probable outcomes of those regimes, to treat them as independent variables. By moving from the nominal/

categorical level of measurement (naming or classifying a regime) to the ordinal level (rank-ordering), an analysis of relations between military style and its various consequences becomes possible. He provided a preliminary rank-ordering of the tendency of the different kinds of military rule to produce political stability and economic development. According to his rank-ordering, the regimes most conducive to 'political" stability are Holding Operations (temporary transitions to civilian rule), after which come Managerial Brokerages (which satisfy group demands outside of the armed forces), Bureaucratic military regimes (where satisfaction of civilian demands is less evident), Personal Dictatorships (with closed power and patronage), and Praetorian systems (with acute instability).

In his multi-dimensional typology, when dimensions changed, his rank-ordering changed. So when military regimes were ranked according to the degree to which they were conducive to "economic" development the Managerial Brokerage was highest (satisfying group demands), then Bureaucratic and Holding Operations (the one because it satisfied group demands less, the other because its time in office limited its achievements), and last Personal Dictatorships and Praetorian regimes (where patronage and corruption are combined with terror and bloodshed), which result in the lowest levels of development (Figure 5.1).

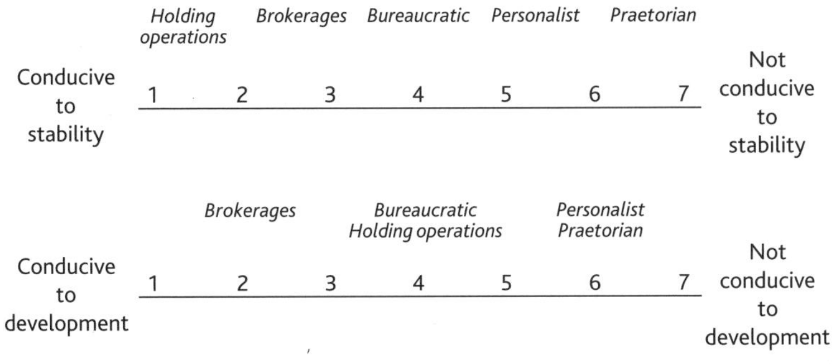

Figure 5.1   Political stability and economic development under military rule

What is extremely interesting about this ranking, developed by someone who was not interested in or even aware of the "oil curse," is that it can be used to challenge one of the most nefarious arguments made about military dictatorships in African oil-exporting countries: i.e. that regime stability is good for development. How many times will apologists for these regimes repeat that the consolidation of power in the hands of a personal dictator is somehow better for development? Yes, mining investments need stability, and, yes, personal military dictatorships are more conducive to stability than are praetorian regimes. But the mere fact that a personal dictator centralizes power in his hands, and thus ends the acute instability and plotting and jockeying for supremacy within the junta, does not mean that his regime will be any more successful at using oil revenues for economic development.

Why not? Problems facing military rulers are numerous. Thomson (2004) has enumerated five common difficulties faced by all military rulers.

First is the problem that any soldier who seizes power by coup d'état establishes a "dangerous precedent." (2004: 138) He who comes to power by coup d'état soon finds he must protect himself from a counter-coup staged by his own soldiers. Therefore one outcome of military rule is an increase in military spending. Soldiers prioritize the corporate interests of the military as an organization; instead of the society as a whole.

Second is the problem that soldiers have little training or experience in running the necessary civilian administrations of modern government. "Coup leaders soon realize that it is much easier to topple a government than it is to build its successor." (ibid.) So they eventually co-opt civilian bureaucrats from the former regime, or at least find competent replacements, if their regime is to accomplish anything but the most rudimentary functions of maintaining order.

Third, as they come to power illegally, by coup d'état, they have a crisis of legitimacy. To remain in power, military rulers must build some kind of linkages with the society over whom they rule. Since these linkages, in Africa, tend to be based on patron–client networks, even military rulers who come to power promising to combat corruption will find they have to use traditional patronage to establish their legitimacy. Fourth, since they always come to power promising to improve on the previous regime, they create a problem of rising expectations. Having promised to improve the economic well-being of the people, soldiers nevertheless find they suffer the very same structural constraints which burdened their civilian predecessors. Fighting corruption and developing the economy is easier said than done. "Effectively, a successful military administration has to produce more from the same resources held by the previous regime." (2004: 139)

Finally, the soldiers who justify their coup d'état by criticizing the abuses of corrupt and inefficient civilian governments do not usually have a blueprint for a better society. Even if they do have a viable economic plan, they often do not have the time necessary to implement structural reforms. "If they are to form a caretaker administration, in order to build the right environment for a just society, how long should they stay in power?" (ibid.) One major dilemma for any military regime is that if it keeps its promise to restore civilian rule, the next government may not pursue reforms. But if it doesn't keep its promise, and remains in power instead, then it will continue to suffer from the other four problems outlined above. Most coups were instigated in Africa without having a social or economic plan. Once in power, soldiers must find new answers to the same old problems, and they must find them rapidly. "Again, it is easier to overthrow a bad government than it is to create a good one." (ibid.)

Map 5.1   Congo-Brazzaville

THE MILITARY IN CONGO-BRAZZAVILLE

The Republic of Congo we see today on the map (not to be confused with Congo-Kinshasa) was born in 1886 out of the separation of "French Congo" (Gabon and Moyen-Congo) under the French commissariat general Pierre Savorgnan da Brazza. At first its capital was located on the Atlantic at the port city Pointe-Noire in the normal pattern of an export-oriented colony. But the scramble into the Congolese basin demonstrated the strategic value of Stanley Pool, where the River Congo became navigable, and from where all of the French colonies in equatorial Africa could be governed. So the Congolese colonial government was moved in 1904 to its present location at Brazzaville,which four years later became the capital of a vast Equatorial African Federation (today's Congo, Gabon, Central African Republic (CAR), and Chad). In the 1920s a railroad was constructed linking Brazzaville to Pointe-Noire, giving Congo its present incarnation as a transit state.

The history of the Congolese army is intimately linked to the history of French colonial troops, whose penetration into the Congo was made on the basis of bloody wars opposing the colonizers and the native peoples. Those Africans who resisted had to defend themselves with makeshift arms such as flintlocks, muskets, throwing knives, assegai spears, and poisoned arrows. The European soldiers, on the other side, used an armada of light artillery, notably 80 mm mortars, heavy artillery, mountain batteries, 65 mm pieces, as well as modern rifles with repeating carriages. (Lebi 2009: 74) Indispensable to this military conquest were African forces known as the *tirailleurs sénégalais* (so named in 1900, although they came from throughout the French African Empire). Trained by metropolitan officers, these blacks were the vehicle of French penetration into Africa, and the principal force for maintaining the white colonial order. (Challaye 2003)

In 1902 the French occupation of the Congo Basin was limited to a few points. A small number of outposts had been established along the banks of the Congo River. It was the granting of vast territorially defined concessions to a series of large private investors called "concessionaires" who were the effective colonial presence in the forests in those early days. They hired private militias to round up local villagers who were then used as forced labor. Women and children were held hostage in the villages while their men were forced to bring back rubber (often caked on their naked bodies) or haul timber in exchange for their families. The abuses practiced by these unscrupulous concessionaires became infamous in Europe as the "Red Rubber Trade." They provoked revolts by exploited Congolese that private militias paid by the concessionaires proved inadequate to quell. This was how the colonial armies arrived, to suppress native revolts, and enable forced labor by profiteering French companies under the "Concessionaire Regime." (Coquery-Vidrovitch 1972)

Using military archives and colonial manuscripts in Paris, Colonel Simplice Euloge Lebi spent a decade meticulously documenting what can be known about the details of the penetration of colonial troops into his country: the 3rd Company's expedition into Ngala region (1902–13), where the people had abandoned their villages to avoid paying hut taxes to the Compagnie Française du Haut Congo (a legal pretext to force moneyless peasants into forced labor); the 1st and 5th Companies' battles against the Kongo peoples (1907–13) living between Brazzaville and Pointe Noire, allowing them to be enslaved in the construction of the Congo-Ocean Railroad; and the 2nd Company's conquest of the "indomitable" Téké (1913–14), who lived in and along the border with Gabon. With modern French rifles that could be loaded and reloaded an estimated ten times faster than flintlocks, the colonial army defeated all resistance and deposed all traditional rulers. (Lebi 2009: 58–75)

Recruitment of black soldiers into the white-officered colonial army, and use of those blacks to commit what today would be called "crimes of war" against their fellow Africans, led to an image of the army as a brutal, collaborating, oppressive agent of foreign domination. At first the *tirailleurs sénégalais* were imported from Senegal, Morocco, and Madagascar, which made it easier for them to kill their African brothers. But it was during the First World War, when France realized the enormous pool of potential soldiers in equatorial Africa could be used to fight their European enemies that the conscription of Congolese populations began. In 1912, before the outbreak of the First World War, there were only 16,000 African soldiers in its colonial armies. Wartime conscription accelerated at such a pace that during 1915–16 the metropole was able to conscript 50,000 Africans in one month alone. (M'Bokolo 1992: 333) In total, it has been estimated that France mobilized 600,000 Africans and Indochinese to fight for them in the First World War. (Lebi: 79)

Black African soldiers were later used to defend France from Nazi Germany in 1939, and in the Armistice of 1940 the defeated Vichy government of Marechal Pétain was authorized to maintain

70,000 men in Africa to protect the Empire. These native colonial troops, along with others recruited by the Free French Forces during the Second World War, were later used by General Charles de Gaulle in the Liberation of France in 1944. In fact, it has been estimated that one-half of the 1st Army that landed on the beaches of Normandy and the South of France to drive out the Germans, that is 125,000 out of 250,000 troops, were of African origins. (Deroo and Champeaux 2006: 199)

Brazzaville itself had served as the capital of Free France during the German occupation, and thus gave the Congo its military vocation. After the war the *Bataillon de tirailleurs de Brazzaville* was created in 1953, commanded by 43 European non-commissioned officers, with 25 African non-commissioned officers to assist them in leading 577 African troops. (Lebi 2009: 100) Recruitment of these first troops had been done in the two major cities, Brazzaville and Pointe-Noire. It was less expensive—the federation capital was teeming with youths. Assimilated urban Africans had received more education. (ibid.: 101) But a change in recruitment policy after 1945 started to select Africans from poorer rural areas, seeking out the most warlike ethnic groups, but above all soldiers from those northern ethnic groups who presented less of a challenge to French domination. Because the pre-war colonial system had privileged the coastal Lari and the assimilated Kongo peoples of the South, it became clear to French colonizers that it was preferable to recruit soldiers from poor rural northerners. Northern recruits were not involved in any of the nascent anti-colonial movements, and more important, they could be used against southern agitators in a classic strategy of divide-and-rule.

When political independence was granted to the Republic of Congo in 1960, there were two kinds of soldiers in the army. The first were ex-colonial soldiers, who had received their training in colonial adventures and the two world wars. The second were newer recruits, including young officers who had received their training in France. The former suffered from a perception of being collaborators, and the low status that had been accorded them by the white colonial officers. There was also a difference in ethnic composition, as the latter had been largely recruited from northerners, who found themselves under southern officers in the new national army. As a consequence, the French colonial legacy not only left behind a low image of the army as a collaborating force, but also sowed the seeds of ethno-regional, north–south divisions within the military organization itself.

The 1st *Compagnie Congolaise* was created in 1960, with three officers, 13 non-commissioned officers, 30 corporals and 102 foot soldiers. (Lebi 2009: 115) Among the non-commissioned officers in the official 1960 listing appear two names of particular importance to the future military regimes: Sergeant Marien Ngouabi, a northern Mbochi who had just received parachute training from Egyptians, and Sergeant Joachim Yhombi-Opango, another northern Mbochi. They came from the same poor, rural northern region, Likouala-Mossaka, and they had both been trained by the French at the prestigious St-Cyr military academy. (Lebi 2009: 104) In 1961 the 1st Company was renamed the *Forces Armées Congolaises* (FAC), in order to emphasize its national mission over its neocolonial French logistical reality. Those older officers who had been formed by the colonial army formed a small tightly knit southern minority who held all the top posts. Beneath them were the younger northern officers, who had not experienced war, who shared regional and ethnic identities that bound them together, and who had received advanced training in France. One of these young officers was Lieutenant Denis Sassou Nguesso, a northerner who had attended the infantry training school in France and had been integrated into the army paratroopers. We will speak of him momentarily.

President Fulbert Youlou (1960–63), a southern Lari priest, was suspicious of the northern dominated army corps, so he chose to rely on the new Gendarmerie Nationale instead. The gendarmerie, usually a kind of military police, was transformed into his presidential guard, reflecting simmering ethnic divisions, and setting a precedent for future praetorian regimes. In addition to this sociological development in the military, another far more dangerous phenomenon was brewing in civil society. Brazzaville had been the federation capital, but after independence it was reduced to the capital of a country with some three million inhabitants. Youlou inherited a "bloated, underworked, inefficient, corrupt, urban-based bureaucracy." (Decalo 1976: 134) Brazzaville contained the bulk of the country's population, and that population was demographically young. Congo's youth exerted a dominant influence in politics after independence, a floating pool of unmobilized rootless manpower searching restlessly for a political role as well as jobs. Trying to employ this restless urban mass in an overstaffed civil service, and presiding over an extractive economy run by the French, Youlou's administration was notable for its increasing corruption, as well as his overt personal contempt for Congo's youth. (Ballard 1966)

President Youlou was overthrown in 1963 by young urban masses in three days of demonstrations and strikes that are today known as "*Les Trois Glorieuses*" (August 13–15) which "were neither a revolution—as they are viewed in Brazzaville—nor a classical coup by Congo's then minute, largely French-officered army." (Decalo 1976: 139) Technically this was not an army coup. The Mbochidominated army simply withdrew its support from the regime, thus enabling its downfall. It did not impose an army officer as an interim president of a caretaking holding operation. Instead the soldiers transferred power to a civilian unionist named Alphone Massemba-Debat, a schoolteacher by profession with a socialist ideology who declared that Congo was henceforth a Marxist–Leninist republic.

President Massemba-Debat (1963–68) proved unable to control the rising expectations he inspired in the country, and in particular the revolutionary forces he unleashed within the Congolese youth, who quickly gained control of his party, the *Mouvement National de la Revolution* (MNR) and created a parallel "revolutionary" army that enjoyed all the favors of the government. The youth wing of the MNR made the official armed forces of FAC seem like a poor cousin of the party's new revolutionary militia, the Civil Defense Forces.

So Massemba-Debat was overthrown in another coup d'état in 1968, this time led by a northerner, Captain Marien Ngouabi (1968–77), the first of three northern, military rulers. Ngouabi was widely respected by ordinary Congolese for his intelligence, incorruptibility, and work ethic. One of his first acts in office was to demand that the Civil Defense Forces hand over their arms. When certain diehard members of the militia refused, he attacked them with the regular armed forces, killing hundreds and effectively returning to the army its monopoly of the use of violence. In addition to founding a one-party Marxist–Leninist regime under the Congolese Labor Party (PCT), Ngouabi also disbanded the Gendarmerie, and replaced the largely discredited FAC with a new National Popular Army, which incorporated the armed youth militia of the MNR. He promulgated a new constitution in 1970 and changed the name of the country to "People's Republic of the Congo." It was the era of the Cold War, a period of intense ideological, political, and military rivalry between two superpowers following a balance-of-power logic whereby each attempted to deny the other access to any supposedly "vacant" space, and African states attempted to maximize their diplomatic advantage.

Very few francophone African states remained non-aligned during the Cold War, and on the basis of an elaborate index,

Hveem and Willetts have demonstrated that from 1964 to 1970 Congo-Brazzaville was diplomatically and militarily aligned with the Soviet Union, although economically it remained more aligned with France. (Willetts 1978: 116–27) Socialism in francophone African states should be viewed in its proper context: "While such states openly adhered to the Marxist-Leninist ideology and proclaimed their resolve to build socialism in their countries, their external economic relations were in fact heavily Western-oriented." (Martin 2002: 117) Rhetorical flourishes aside, Ngouabi's ability to create a socialist paradise out of a mineral-exporting postcolonial African state were severely limited. The first oil shock in 1973 did briefly inject windfall profits into the treasury, which allowed him to nationalize a large portion of the country's industry, and to create new socialist state enterprises. But once oil prices fell, his ability to deliver state-led economic development proved a chimera.

One consequence of the new ideological orientation of the regime was the politicization of the armed forces. "Our immediate task," explained Ngouabi, "is to organize the party and, at our level, to implant it in the army so that it can truly run it." (Lebi 2009: 144) This had the effect of introducing a third fracture in the armed forces. First there were generational differences between young and old. Second there were ethno-regional differences between northerners and southerners. Now there were ideological differences between Marxists and anyone else (called "capitalists" or "imperialists"). But more dangerous than this, politicization of the armed forces removed the barriers between the military and civilian functions of government. "Indeed," lamented Decalo, "under military rule the pace and intensity of disorder has been heightened because of the total politicization of the armed forces." (Decalo 1976: 168) After seeing his popularity decline as oil prices fell, Ngouabi was murdered by gunmen in a dramatic shoot-out outside his residence in 1977. His successor was General Joachim Yhombi-Opango (1977–79), a fellow northerner from the same region as Ngouabi, but who did not share his predecessor's Marxist ideology. Yhombi-Opango was a military man, not a politician, and in order to restore the relative power of the army over the party, he established a Military Commission in the PCT that became the true center of power.

Seated on this Military Commission was Colonel Denis Sassou-Nguesso, another northerner, born in 1943 in the region of Oyo. Today president, he is the main actor in this case study. Sassou came from Mbochi parents. His father was a hunter and tracker, and one of the clan chiefs in the village of Edou, who initiated his son as a *mwené* (clan chief) when he was only five, which privileged him to special education and preparation. His schoolteachers in the village encouraged his father to send him off to middle school, a hundred kilometers away, to continue his education. This enabled him to attend one of only four such schools in Congo. He studied five years at middle school at Dolisie, in the south, where he was mentored by a French teacher who later encouraged him to take an entrance exam organized by the colonial government to train and commission black officers for the new national army. (Sassou-Nguesso 2009: 25–35) He was one of the young northerners recruited by the French to counter the rise of southern elites in the twilight of colonial rule.

Sassou received military training in an accelerated six-month training course run by the French army in Bouar in the CAR. Then after qualifying as a paratrooper he went to officer's school in Algeria in 1961–62. He returned to the Congo with the rank of second lieutenant, but was recruited to round off his military training in France at the Infantry Application School in Saint-Maixent. "Back in Brazzaville I was assigned to work with Captain Marien Ngouabi to found the Congolese First Airborne Group, whose role would be to spearhead our new army." (ibid.: 35) "I was successively commissioned lieutenant, captain, commander of a paratroop corps—an elite unit—for seven years, then Commander of the Military Zone of Brazzaville, Commander of the Land Forces, Director-General of State Security, and finally Minister of Defence from 1975–1979. I held those final responsibilities under President Marien Ngouabi." (ibid.: 35–6) Sassou had been part of the first 1968 coup d'état that brought down the Youlou regime, and was one of the earliest officers to join the PCT in the Ngouabi era. By the time Yhombi-Opango had appointed him to the Military Commission, he was considered a party loyalist, a hard-core Marxist-Leninist, and a survivor. Behind his socialist façade, however, was a man burning with personal ambition. Within two years Yhombi-Opango was removed from power by the Military Committee, and replaced by the 36-year-old Sassou.

The first presidency of Sassou-Nguesso (1979–90) continued to wear the costume of a Marxist–Leninist one-party regime. But in reality it was an oil-rentier state, funded by the French oil company, Elf-Aquitaine, and the Italian national oil company, Agip-Mineraria. Like its southern neighbor the People's Republic of Angola, whose ruling MPLA also preached orthodox Marxist–Leninist ideology, the ruling PCT in the People's Republic of the Congo earned its money selling petroleum exports to capitalist multinational oil corporations. Sassou had support in the Mbochi-dominated army, and, as a hard-line Marxist, he was also highly regarded by politically powerful southern intellectuals. Today he explains his Marxist–Leninism as a pragmatic response to the Cold War: "The only allies of our third-world countries in their liberation struggles at the time were the countries of the Communist bloc [...] That's what determined our choices as future leaders of Africa, far more than strict Marxist ideology." (Sassou-Nguesso 2009: 38) Sassou managed to keep the one-party state going for twelve more years by co-opting or eliminating ethnic, party, and civilian-sector rivals. Despite keeping relations with the Soviet Union, China, Cuba, North Korea, and other communist states, however, France remained his largest trading partner and major source of foreign aid. Elf-Aquitaine remained the chief exploiter of the country's oil, "the production and sale of which accounted for more than 90 percent of Congo's foreign earnings during the 1980s." (Clark 1997: 65)

During his first presidency, Sassou chose his army chief of staff from ethnic northerners: Colonel Raymond Damase Ngollo (1975–82), Colonel Emmanuel Elenga (1982–87), and Colonel Jean-Marie Michel Mokoko (1987–92). But in order to maintain the façade of a one-party state, in the Cold War model of soviet socialism, everywhere an officer exercised his functions, it was the PCT that conferred upon him both his power and his rights. One regime insider, General Norbert Dabira, later confirmed in an interview that "to exercise a leadership role in the army was above all of a political nature." (Lebi 2009: 147) As time progressed, Sassou's ability to keep the regime's socialist promise of material welfare was compromised by heavy reliance on a single commodity, declining terms of trade, and an exploitative relationship with the country's former colonizer. To provide jobs, the civil service swelled to 73,000 in 1987, and alone accounted for 38 percent of the budget. He also purchased consent by giving jobs in the 85 money-losing state enterprises, which, according to the World Bank, accounted for 28 percent of the country's total external debt. (*African Economic and Financial Data* 1986: 171)

Congo's foreign debt, which was equal to its GDP when Sassou first came into office, had increased to nearly twice the size of its GDP by 1990. "The sources of this debt were to be found in the inability of Congo's governments to accept the severe limits that their meager resources put on socialist development." (Clark 1997: 66–7) In December 1990 the World Bank suspended further loans because of non-payment. At this point Elf-Aquitaine agreed to reschedule Congo's debts in exchange for new oil concessions, at bargain basement prices. Elf then provided new loans at exorbitant rates. The rescheduling deal, and its link to the award of the giant Nkossa field to Elf, was later described in a French court by Jacques Sigolet, director of Elf's FIBA bank, who testified that "Elf would set up a company in Switzerland, which would lend at a higher interest rate to a bank, which would then lend at an even higher rate to Congo." (Shaxson 2008: 253) This was the beginning of a Byzantine form of oil-backed loans that would further plunge the country into economic collapse. The economic crisis, however, coincided with the fall of the Berlin Wall, forcing the regime to implement political reforms. In 1990 the PCT abandoned Marxism–Leninism (now that its ideological ally was gone) and under pressure from the French (following the La Baule speech by François Mitterrand) opened elections to multiparty politics. Twenty-two new opposition parties emerged, and, following a general strike called by the Congo's sole legal union, called for a "National Sovereign Conference," which quickly became a 105-day constitutional convention.

The army played an important role in this historical moment of constitutional reform, both protecting the national conference with a military ring around the parliament building, and respecting its decisions as sovereign. When the conference was over, the armed forces declared their official neutrality *vis-à-vis* all political parties (Lebi 2009: 158) and "Congo's highest military commander reaffirmed an earlier pledge that the army would not interfere in the political process." (Clark 1997: 68) At the end of the conference the delegates chose an interim government headed by a former World Bank official, André Milongo, with the title of "prime minister." Meanwhile the presidency, retained by Sassou, was reduced to a largely ceremonial position.

A military coup was one of the principal challenges that the new government faced during the year of transition. Several members of Milongo's cabinet from the Pool region had plotted to bring the army under political control by spreading rumors of a coup. Milongo also appointed General Michel Gangou, a sworn enemy of Sassou loyalists in the army, as the new military chief of staff, and then began purging northerners from the military high command. "This led to a tense stand-off between Milongo's government and the army, which at one point involved a shoot-out causing six deaths and Milongo's seeking asylum at the American embassy." (Clark

1997: 69) Who was behind this attempted coup? "Those who were advocating democratic debate were unable to comply with its rules," claims Sassou. "It was then that the violence began, with Lissouba and even Milongo, since both refused to honour the principles they had set for themselves." (2009: 50) But according to Milongo, the real cause behind the attempted coup was oil, and the corrupt practices of Elf in Congo.

> It was a coup to get rid of me. It was the army, and behind them, Elf! A senior Elf man flew in on an Elf plane to organize it. [...] They wanted to get rid of me to stop the political process. They did not want the changes because Congo would have discovered their secrets. (Shaxson 2008: 109)

FROM PRAETORIAN RULE TO PERSONAL DICTATORSHIP

What should have been a happy ending to military rule turned into ethnic civil war. Two southern civilian politicians, Bernard Kolélas and Pascal Lissouba, soon rose to prominence in the legislative and presidential elections of 1992. But the ethno-regional character of those elections can be seen in their final results. Kolélas won the Pool region and those parts of Brazzaville where the Lari and his Bakongo peoples predominated. Pascal Lissouba, a Njabi, won the southern vote in Niari, Bouenza, and Lékouma. Sassou won the northern regions of Cuvette, Sangha, Likoula, and Plateaux. Since the northern regions were the least populated, Sassou was eliminated in the first round, leaving Kolélas and Lissouba to fight it out in the second round. Lissouba won, and in 1992 became the first democratically elected president of Congo since Youlou. But divisions between Lissouba and Kolélas quickly turned into a vote of no confidence that required new legislative elections to be held in 1993. Accusations of vote rigging led Kolélas to withdraw from the second round, and both sides began acquiring arms. There were a number of deaths in this period as the capital became the scene of numerous skirmishes among the Congolese militia and a variety of armed political forces representing Lissouba, Kolélas, and Sassou.

> Brazzaville had crystallized into fiefdoms held by militias— Cobras, Ninjas, Cocoyes, Mambas, Aubevillois, Zulus, and more—and tanks were rolling in the streets. In the ensuing fighting perhaps 2,000 people died, and 300,000 people fled, as drug-addled thugs raped, decapitated, and skewered their

way through Brazzaville suburbs [...] guns flooded in and were dispersed among Brazzaville's households. (Shaxson 2008: 112)

The military did not stage a coup d'état. Sassou left Oyo for a mansion outside Paris, beginning a three-year period of self-imposed exile (1994–97), where he offered his services to French businessmen, and enjoyed a lavish lifestyle that was difficult to explain for a former Marxist–Leninist dictator. He sold himself to the French as the only person capable of restoring peace to the Congo, and worked closely with Elf's *monsieur Afrique*, André Tarallo, to orchestrate a return to power. He created with his personal fortune a 700-person private militia called the "Cobras," who stole a large number of arms and munitions from the army training school at Gambona, near his hometown, and surrounded his Oyo estate. His "Cobra" militia warded off attacks in the Brazzaville districts of Ouenze, Mpila, and Talangai inhabited by northerners, while Sassou forged a new coalition of northern-based parties in a deeply divided legislature. Meanwhile President Lissouba, fearing his armed rivals, abused his presidential powers. He ordered the army to blockade the Bakongo section of Brazzaville, from which Kolélas "Ninjas" were operating. For the first time the army used heavy weapons, causing major collateral damage in the "Ninja" stronghold.

Although it is common to date the civil war to 1997, two Swedish scholars doing research in Brazzaville reported widespread ethnic violence by 1994:

The victims were burned, buried alive, shot, thrown into the river, decapitated and/or slashed with machetes. Among the victims were men, women and children [...] Women and very young girls, sometimes mothers and daughters, were gang raped. (Clark 1997: 74)

Another coup plot was uncovered and there were army mutinies as a rampage by Sassou's "Cobras" threatened to topple the regime. President Lissouba began to get intelligence about arms flown in from neighboring Gabon to Sassou's home in northern Congo. (President Bongo had recently married Sassou's daughter Edith, so Sassou was now his father-in-law, despite having ethnic ties with Lissouba.) Suspecting that Elf oil interests were behind Bongo's behavior, Lissouba banned public protests and paid Israelis to train his militia. (Shaxon 2008: 113) When Sassou returned to Congo in 1997 and declared his candidacy in the presidential elections,

Lissouba called out the armed forces. Early in the morning of June 5, 1997, eight armored cars surrounded Sassou's compound in Oyo. The "Cobras" fired a rocket, which hit the nearest vehicle, and fighters opened up with AK-47s. In the official history, this was the outbreak of the civil war: "Lissouba intentionally caused the incident because he was afraid of losing the election." (Sassou 2009: 72)

But another unofficial version is that Sassou's return to power had been orchestrated by the French, in particular, by Elf Aquitaine. According to this version, France was concerned about the rise of the English-speaking Kabila regime in neighboring ex-Zaire, and Elf was concerned about losing its oil concessions to the Americans (following a deal reached by Lissouba with Occidental Petroleum) and also feared revelations of its corrupt practices. There are so many contradictory stories published about the civil war that it is difficult to know who is telling the truth. We know, however, that Pascal Lissouba unwisely made an alliance with Angolan UNITA rebels, which provoked a reaction from the ruling MPLA in Luanda (a long-time ally of the military regime in Brazzaville). So, to cut a long story short, in 1997 Sassou's "Cobras" captured Lissouba's suburban strongholds in Brazzaville, took the airport, and then took the presidential palace itself, while the Angolan MPLA sent an armored column, backed by fighter bombers, into Lissouba's Pointe-Noire. "Angola's disciplined, battle-hardened, and better resourced soldiers easily overran Lissouba's rabble, and Sassou was soon back in power." (Shaxson 2008: 114)

Ten thousand people reportedly died in that year of fighting. But the exact number will never be known. French President Jacques Chirac, who was a personal friend of Sassou, welcomed the return of peace and stability to Congo. The French mainstream media propagated the idea that the civil war had come to an end. Elf-Aquitaine preserved its interests in the country. Although vocal critics in the alternative French media condemned Sassou's return to power, and independent French judges indicted Sassou and several members of his government for the execution and disappearance of 350 returning Congolese refugees (the "Beach" massacre of May 1999), the regime denied the charges, which were soon dropped, and relations between Paris and Brazzaville were normalized. A civilianized Sassou was by then wearing elegant suits (tailored in Paris) and presenting himself again as candidate in 1999 presidential elections. After three years of intrigue and delay, Lissouba and Kolélas boycotted these elections when they were finally held in

2002. Sassou took 90 percent of the vote, a notable improvement on his 17 percent in 1992. Southern-based rebels, realizing that they had no foreign allies to help them, finally agreed to a peace accord in 2003. However, Kolelas' "Ninjas" remain active, and continue to camp in the jungle on the outskirts of Brazzaville, where they live on banditry and smuggling to make ends meet.

To accept this masquerade of presidential elections as a return to democracy is nothing short of naïve. Sassou's recent re-election in 2009 was neither free nor fair. The "civilianized" military regime of Congo-Brazzaville must be recognized for what it is. However, the purpose of this case study is to classify Sassou's style of military rule in order to explain its effects on the oil curse. Has Sassou's style of rule created political stability or economic development?

It is tempting to describe the Congolese military regime as Praetorian, which was how Decalo described it. The continued presence of parallel armed militias in the Congo today would suggest a system "typified by intense intra- and inter-elite strife, the presence of continuing jockeying for supremacy within the ruling junta, and a perennial tug-of-war for influence and power between various groups and military factions." (Decalo 1976: 243) The distinguishing feature of praetorian rule is acute instability coupled with low achievements in the socio-economic and political arenas. But the Congolese army is no longer in control of the Congo—rather the regime has devolved into a Personal Dictatorship under Sassou. Power has been "centralized into the hands of one leader, who uses selective terror and purges to maintain himself in office." (ibid.: 245)

"Personal Dictatorship" is the ultimate in decay, and least likely to restore civilian rule. Those who think anything is better than war can be satisfied by such peace without justice. But since the end of the civil war, peace and stability have not provided the foundation for economic development. The effects of Personal Dictatorship on the economy and society of Congo have been wholesale theft of the country's oil revenues from a deeply impoverished population that is ranked among the lowest in the world in human development. It is difficult to imagine that Congo-Brazzaville once had one of the highest per capita incomes in sub-Saharan Africa, and that its population enjoyed one of the continent's highest literacy rates. Large oil revenues and a small population promised to achieve the regime's socialist ideals of a better future. Where did all the oil money go? Scandalous revelations of the regime's use of oil-backed loans and its corrupt practices of offshore banking, by Global Witness (2004), suggest that Congolese oil revenues have been used for personal enrichment of Sassou, his family, and those who benefit from his personal patronage. If during Sassou's first Praetorian regime the Congolese government borrowed money to fund a socialist program, during his second Personal Dictatorship he has borrowed money to amass personal and family fortune. Sassou and his collaborators have literally stolen his country's future oil revenue through oil-backed loans. "It's untrue, completely untrue!" claims Sassou (2009: 183), but the charges are not without foundation.

Much of what we know about the systematic corruption of the Sassou regime comes out of a trial of 37 former senior executives of the now-defunct Elf that ended in Paris in 2003. A 600-page indictment listed allegations of corrupt behavior by top Elf officials including siphoning off commissions into secret banks accounts, buying multi-million dollar properties and expensive jewelry, and embezzling money. André Tarallo, Elf's director for Africa and of hydrocarbons, was eventually sentenced to four years in prison (which he never served) and fined $2.5 million (a pittance considering the sums involved). The real importance of the Elf affair was the information it revealed about how the company paid African leaders from offshore bank accounts to maintain its powerful market position. The prosecution provided extensive details about how offshore structures were set up to buy off African leaders, by diverting signature bonuses (one-off payments) and subscriptions (money skimmed from oil sales) into personal bank accounts. It also revealed how African leaders were encouraged to take out oil-backed loans that enriched the company's secret funds. "Bonus payments" were the most basic element of the system. According to Tarallo's testimony, they usually ranged from $1 to $5 million, but sometimes exceeded $10 million. These were paid directly to the African leaders or their subordinates, and rarely appeared in public finances. "Subscriptions" were ongoing payments tied to oil sales that were funded by Elf Trading, by under-invoicing crude oil it bought from its subsidiary Elf-Congo and then selling that oil with an average mark-up of 20 cents per barrel, and placing the money with an additional 20 cents per barrel in secret Elf trusts located in Liechtenstein. (Global Witness 2004: 23)

Finally, "oil-backed loans" were ostensibly granted to allow the Congo to pay its civil servants salaries and avoid revolts, but as Jacques Sigolet, Elf's financial officer who handed these accounts later confessed, these loans had been "conceived in such a way that the Africans were only aware of the official lending bank and were ignorant of the whole system which Elf rendered particularly and deliberately opaque." (ibid.) A portion of each loan was held back as "syndication rights," which, according to Sigolet, went to Sassou and other state authorities in Elf-controlled bank accounts in Switzerland or Liechtenstein. Often money was deposited in bank accounts held by Sassou in the Banque Française Intercontinentale (FIBA), a bank created by Gabonese president Omar Bongo in 1975, with Sigolet serving as its president for two decades, until 1996.

In conclusion, this case study provides evidence that soldiers are not *per se* any more capable than civilians of defending their resources from major foreign powers, resisting multinationals, adopting international good governance initiatives, or combating the negative effects of oil rents. Sassou-Nguesso has financed his Personal Dictatorship with oil-backed loans, and in the process, literally mortgaged his country's future. He has collaborated with foreign powers, in particular France, becoming an agent of multinational oil corporations. He failed to implement good governance initiatives, despite signing them, and on the contrary continued to channel his country's oil revenues into offshore bank accounts that have provided him with a huge unaccountable revenue stream. Finally, after three decades in power, his military regime has failed to combat the negative effects of the oil curse. It is easier to overthrow a bad regime than it is to create a good one. People should not await their salvation from "the man on horseback." To change the condition of the Congolese people, action must come from below the level of the state.

Part II

Power From Below

6
Journalists and Intellectuals

THE TREASON OF THE CLERKS

One reason African oil regimes are so despotic, corrupt, and violent is that no one stops them. As the preceding chapters have shown, those key external actors who have the power to change the situation, such as the major foreign powers who import African oil, or the large corporations who export it from Africa, continue to do business with despots. They provide them with money, arms, and complicit support. Potential foreign sources of positive change such as international good governance initiatives, while necessary, have proven to be insufficient, since they rely on the oil regimes to change their own behavior yet lack the strategic military power of states and the financial–technical clout of corporations. Consequently, the outside world lets ordinary Africans suffer the abuses of kleptocratic praetorians, patrimonial autocrats, and police states that are incapable, or unwilling, to change themselves. So the people who live in these dictatorships should not count on salvation coming from an international community that seems to prefer African resources to African people. Nor should they count on their rulers to reform themselves, given the corrupting influences of oil, accumulation, and power. Real change will come not from above, but from below.

In order to overcome the oil curse, one must first recognize that it is a problem. If international good governance initiatives exist at all, it is because the problems of oil have been recognized. This is the first step. If domestic efforts to fight the oil curse exist at all, it is because Africans recognize that it *is* a problem. Consciousness is necessary before action, and correct thinking is necessary for correct action.

This chapter concerns itself with "cognitive liberation" (McAdam 1982) generated by writers who provide the basis for the *transformation of consciousness* essential for Africans to mobilize against their oppressors, both domestic and foreign. Their writings take the form of "journalism" (e.g. investigative reports, opinion-editorials), "scholarship" (e.g. empirical research, conference papers, scholarly monographs, and edited volumes) and "literature" (e.g. novels, poems, and lyrics). Africa has a long tradition of writers who exposed earlier unjust forms of trade. Narratives by former slaves such as Oulada Equiano that exposed the evils of the slave trade in the eighteenth century became a source for abolitionists in the nineteenth century. Critical writings of men such as Franz Fanon that exposed the evils of colonialism in the first half of the twentieth century provided inspiration for the anti-colonial struggles of the second half. African intelligentsia can provide inspiration today for activists fighting the oil curse.

The first problem for African writers is their marketplace. Africa is poor, and most Africans can't afford books. So African writers find that if they want to become famous or recognized internationally as "intellectuals," they will eventually have to be published by foreign publishers, promoted by foreign mass media, discussed by foreign think-tanks, taught in foreign universities, and reviewed by foreign scholars and critics—in short, "legitimized" by institutions outside of Africa. The absence of large paying audiences for African writers in Africa means that to succeed, they must publish with European or American publishing houses and so must write for northern audiences. This creates a dilemma. To succeed, they must please their former (and current) oppressors, and they are obliged to create in the language, and adopt the mental framework, of those who jealously guard the privileges of cultural domination. Black writers who defy them are criticized, marginalized, or ignored.

The second problem facing African writers is their governments, particularly the corrupt, violent oil dictatorships. There is little civil liberty and even less freedom of the press in most African oil states (see Table 6.1). Journalists who investigate corruption, and editors who publish this information, are systematically menaced, arrested, imprisoned, or killed. Africans who dare to expose their regime from inside write in a climate of state censorship, intimidation and repression. It is by no means uncommon for newspapers in the African oil regimes to be shut down, their printing presses destroyed, their offices sacked, and computers seized. Reporters have their passports taken, their families assaulted, and their persons battered. How many brave journalists have ended up as corpses floating in the river? How many reporters have been tortured to death in prisons? How many have ended mysteriously "missing" or reportedly committing "suicide" in their cells? Visit the sites of Amnesty International (AI) or Reporters Without Borders (RSF) and you will find a catalogue of government efforts to silence those dissidents who speak truth to power.

Table 6.1   Civil liberties and press freedom in African oil states

	Civil Liberties[1]	Status	Press Freedom[2]	Status
Angola	5	Not Free	61	Not Free
Cameroon	6	Not Free	65	Not Free
Chad	6	Not Free	76	Not Free
Congo	5	Not Free	53	Partly Free
Eq. Guinea	7	Not Free	90	Not Free
Gabon	4	Partly Free	69	Not Free
Nigeria	4	Partly Free	54	Partly Free
Sao Tomé	2	Free	28	Free
Sudan	7	Not Free	78	Not Free
Average	5.1		63.8	

1   Civil Liberties rank – ordered: 1–2.5 (Free) 3–5 (Partly Free) 5.5–7 (Not Free).
2   Press Freedom rank – ordered: 0–30 (Free) 31–60 (Partly Free) 61–100 (Not Free).

Freedom House, *Freedom in the World Survey* (2009), *Freedom of the Press Survey* (2009) www.freedomhouse.org.

The third problem facing the African writer is the absence of public institutions conducive to his intellectual output. Most international efforts focus on primary and/or secondary education in Africa; few concentrate on higher education. Public universities in African oil states are often worse than those in non-oil states. Journalists and intellectuals are potentially dangerous to corrupt oil-rentiers who, instead of investing in the highest human development of their societies, embezzle revenues for personal enrichment, allocate resources to purchase political support or arms, and waste public money on prestige projects (and the conspicuous consumption of luxury goods). Since their universities are dilapidated, African intellectuals seek higher education abroad, usually in former colonial powers, where the rich architecture of intellectual life is available, and where they can write fluently in the imperial language. Some African intellectuals remain expatriates—you meet them as black African professors in your universities, often second-class citizens of academia. Others return to their countries of origin to become irregularly paid teachers in impoverished and ill-equipped classrooms, with few resources to launch a world-class intellectual career. When their country of origin is a police state, they are censored and isolated.

The fourth problem facing the African writer is the collaboration of other African writers. The emergence of a social movement requires a *transformation of consciousness* within a significant segment of the aggrieved population. Before collective action becomes possible, *people must collectively define their situation as unjust* and as subject to change through collective action. Often what prevents this from occurring is the betrayal of a certain category of African intellectual. Let us call them "sell-outs." They are spokesmen for the oppressors. Given the first three problems, it is not hard to understand why these writers adapt to meet the demands of their marketplace, their regimes, and the dominant paradigms of the outside world. Nevertheless there is a responsibility of African intellectuals to remain politically engaged on behalf of their people. When the educated elites of African oil regimes speak power to truth, instead of speaking truth to power, and they seek their own advancement to the detriment of the liberation of their race, they perform a great evil. Of all the acts of heroism, one of

---

Box 6.1   The treason of the intellectuals

French essayist Julien Benda sharply attacked the European intellectual establishment of the 1920s for abandoning disinterested intellectual activity and for allowing its talents to be used for political and nationalistic ends. Benda charged that by providing class, race, and national passions with a network of doctrines, giving them moral, intellectual, and even mystical authority, the intellectuals were stirring up hatred and strife, and if fact had become promoters of war:

> Let me recapitulate the causes for this change in the "clerks:" The imposition of political interests on all men without exception; the growth of consistency in matters apt to feed realist passions; the desire and the possibility for men of letters to play a political part; the need in the interests of their own fame for them to play the game of a class which is daily becoming more anxious; the increasing tendency of the 'clerks' to become bourgeois and to take on the vanities of that class; the perfecting of their Romanticism; the decline of their knowledge of antiquity and of their intellectual discipline." (pp. 17–7).

Julien Benda, *La Trahison des Clercs* (1927)

---

the most sublime is creating ideas that liberate Africans mentally from an internalized oppression. Of all the forms of oppression, one of the most insidious is the collaboration of educated elites who teach the people that they are not really oppressed. There are two kinds of African intellectual: liberators and collaborators. The latter perform an African version of what was once described, in another context, as "the treason of the intellectuals."

In his famous "Parable of the Cave" (*Republic*, VII), Plato describes men who dwell in a sort of subterranean cave, fettered from childhood to remain in the same spot, able only to look forward, forced to watch shadows on the cave walls cast by puppets manipulated by their captors. They confuse the shadows on the wall with reality, and "in every way such prisoners deem reality to be nothing else than the shadows of the artificial objects" (p. 123). If one of them were freed from his fetters and compelled to stand up suddenly and turn his head around and to lift up his eyes to the light, at first he would feel pain. Because of the dazzle and glitter of the light, he would be unable to discern the objects whose shadows he formerly saw. "What do you suppose would be his answer if someone told him that what he had seen before was all a cheat and an illusion?" (ibid.: 124–5) If someone were to drag this prisoner out of the cave and into the light of the sun, at first his eyes would be so affected that he would not be able to see even one of the things we call real.

At first he would discern shadows, and after that, the likeness or reflections in water of things, and from these he would go on to contemplate the appearances in the heavens and heaven itself, more easily by night, looking at the light of the stars and the moon, than by day. Finally he would be able to look upon the sun itself and see its true nature. At this point, if he recalled the cave and what passed for wisdom down there, he would count himself happy to know the true nature of things, and would feel pity for the cave dwellers. And if there had been honors bestowed upon those who best described the shadows on the wall, he would not envy or emulate them, but would prefer to live on earth and see the true nature of things than to endure the cave's darkness and ignorance. He would not, in any way, wish to go back down into the cave. But Plato makes it clear that such a man who has "seen the light" has a moral responsibility to descend back into the cave, to go back down again and contend with the perpetual prisoners before his own eyes become accustomed to the darkness. He would have to expose the shadow images, and the puppet masters. He would have to endure the prisoners' ridicule, their accusations that he had been blinded by the light, and perhaps even be killed by them. But having seen the light, the truth, the good (*agathon*), it would be his moral responsibility to help others to escape. "Down you must go." (ibid.: 143)

Map 6.1   Cameroon

CAMEROON: THE *AGATHON* OF MONGO BETI

Mongo Beti (*né* Alexandre Biyidi-Awala) was born in 1932 in the town of Akometan, Cameroon, 60 miles from the capital city of Youndé, at a time when this country, an ex-German colony, was being administered by France under a League of Nations mandate following the First World War. He was the middle of three sons. His father, Oscar Awala, was a poor agricultural worker who, despite his lack of political rights or social standing, showed a tremendous amount of courage: "I believe that the initial image a child forms of his father is very important, even more so if the father dies when the child is still young. I admired him a lot because my father was extremely courageous. He wasn't big, but each time the colonial agents wanted to come into our village and abuse a woman, my father opposed them, although he had no authority. He was skinny like me. They put him in prison, and shackled him. But he just said 'No, she's a widow, she has kids, or it's my brother's wife!'"(I: 172)*

His father died under mysterious circumstances. "The body of my father was dragged out of the Nyong River, at Mbalmayo, on November 2nd 1939. The last time he had been seen was at a party the previous night. Not only did everyone declare that my father hadn't drunk that night, but more, when he was found, his body bore numerous bruises on the forehead, and there were enough evident traces of a beating by blunt instruments. Manifestly, my father had been thrown in the river after being murdered." (I: 315)

Beti attended primary school at Mbalmayo, and later was admitted to a Catholic seminary at Akono. "My father died when I was only seven years old, and my mother couldn't take care of us all, so at an early age I was sent to a mission boarding school. Cut off from traditional Africa, I started studying the French classics." (I: 165) He didn't like Catholic missionaries, and was eventually kicked out, starting a lifelong contradiction of assimilation and rebellion. His mother was a major influence: "It was to her, I have no doubt, that I owe being a writer. The first story I ever told by the campfire was one my mother taught me. Thanks to her I overcame my shyness and accepted to go before the village when my turn came. I remember it well. I was not even six years old. My father, it was a little before his death, grumbled 'What do you want, my son to go from village to village telling stories, like an *mbomvet*? That's all he needs. My son is going to school. He's going to become an administrator, with an office.'" (I: 396–7)

Beti eventually went to high school, Lycée Leclerc in Mbalmayo, graduated in 1950, and then passed the *baccalaureate* in philosophy and letters in 1951, performing so well that he earned a scholarship to study in France, and was admitted to the Faculté des Lettres at the University of Aix-en-Provence. "When I arrived in France in 1951, I stumbled upon a Richard Wright wave, a magic lantern show for a little African barely out of the colonial bush, studying the Bible at the college of letters in Aix-en-Provence, situated in the old

---

* This chapter has cited the collected writings, interviews, articles, and pamphlets of Beti gathered by Boniface Mongo-Mboussa that were posthumously republished by the prestigious French publishing house Gallimard in three volumes: *Le Rebelle*, Vol. I (2007) Vol. II (2007) and Vol. III (2008), cited by their volume number alone (i.e. I, II and III).

town, above the belfry of the town hall, whose amphitheatre was so dark you could read without being noticed." (I: 316–17) He didn't like French writers, but was enthusiastic about African-American writers, who seemed more confident, and wrote about real people. "I didn't like French literature at all. For French literature, as I had discovered it, was a literature of the bourgeois, of good, intelligent characters, not ordinary people." (I: 165). Nor did he find anything he liked in French about his native Africa, declaring categorically in one of his earliest pieces of literary criticism: "If one makes an exception for books written by explorers and missionaries, whose mentality generally grew old with their authors, there isn't a literary work of quality inspired by black Africa and written in the French language." (I: 30)

"There are some excellent poets," he admitted, "like Aimé Césaire, David Diop, Paul Niger, J. Roumain," but he argued that nobody actually reads poetry: "In reality, the category of writer we are most awaiting are novelists." (I: 32) After finishing his license at Aix-en-Provence, he was admitted to the Sorbonne in Paris. His professional goal was to become a schoolteacher, and for the next six years he researched and wrote a doctoral dissertation on "The Image of Black in Bug-Jargal," a subject that satisfied the ethnological perspective of his superiors, who considered such a topic to be suitable for an African. But his *L'Image du Noire dans Bug-Jargal* (1961) criticized the colonial language, particularly its use of "black" as a pejorative. During these years he started teaching to earn his living. It was also during this time that he started writing fiction, and published his first novel, *Ville cruelle* (1954) under the pseudonym of Eza Boto (inspired by Ezra Pound). Enjoying a little bit of recognition among the handful of people who were beginning to read literature produced by francophone Africans, he started working under Alioune Diop on the path-breaking journal *Présence africaine*. This Paris-based literary magazine was famous for launching the *négritude* movement, associated with such famous writers as Leopold Senghor and Aimé Césaire. It provided a new audience for intellectual discontent among assimilated francophone black Africans living in what was still a colonial context. But Beti was discontented with these francophone African writers themselves.

In 1953 he wrote a scathing review of *L'Enfant noir* by Camara Laye, whom he criticized for pandering to a bourgeois French readership that "demands of the African writer to write about the picturesque, and nothing but the picturesque." (I: 33) He said this successful novel was "a story to fall asleep to standing up, full of harems, serpents, sorcerers, eunuchs, beggars, nababs, negro princelings, golden bracelets, café au lait, mulatto women with generous breasts, village festivals, palm wine ... nothing is missing." (I: 43) He criticized Laye as having pandered to the desires of French readers, selling a colonial myth of an idyllic Africa: "Camera Laye wants to tell us a bedtime story." (I: 43) He labeled such writers as black *clercs* (using Benda's term) who pretend not to take a side in the struggle, "and hide behind sorcerers, grandfather snakes, rites of initiation at nightfall, fish-women and an arsenal of picturesque trash. [...] When the black *clerc* affirms having been initiated by the light of the moon, or having belonged to a lion brotherhood, or having caressed a crocodile," (I: 37) the bourgeoisie applaud. Instead an African writer should talk about "the premier reality of black Africa, I might say the only profound reality: colonialism." (I: 36) Beti called for a new kind of African novel in the French language, one that was realistic.

"The time is not propitious for an authentic African literature. For either the African writer is a realist, and in this case, he risks not being published much. Or if he does manage, the critics will ignore him and the public also." (I: 36) "Black writers cannot even write for a black public. Let's face it. The francophone African novelist, white or black, writes essentially for the metropolitan French public, which explains many things. The French who read novels are the bourgeoisie." (I: 34) They missed colonialism, he argued, and wanted to relive a fantasy of domination in their fiction.

The work that brought him his first real fame was *Le Pauvre Christ de Bomba* (1956), which he published under a second pseudonym, Mongo Beti. "When I started writing in the 1950s," he later explained, "my real name, Alexandre Biyidi-Awala, was just too difficult to pronounce, especially on the telephone, where it always had to be spelled out. In 1953 I had known someone who really admired the American poet Ezra Pound, and on that model I had invented Eza Boto. But I was disappointed by the printing quality of the first edition of *Ville cruelle*, and anyway I was angry with my first publisher, Alioune Diop, at *Présence africaine*. So I drew a cross through this book and decided to find myself another pseudonym. I opted for 'Mongo Beti,' which means 'Son of the Beti.'" (II: 13)

The subject of *Le Pauvre Christ de Bomba* was the colonial mission in Africa. It tells the story of a Catholic priest who establishes a mission, tries to suppress the cultural practices of the Africans, but in the end discovers to his dismay that the young girls brought into his mission are used as prostitutes. The style of his novel was realistic. It became a critical success, not only because of this new style, but also because of its exposé of ill-understood practices of missionaries in colonial Africa. This of course scandalized the French Catholic establishment, which denounced the novel's allegations as defamatory. The next year he published *Mission terminé* (1957), which won the Prix Sainte-Beuve. Despite their critical success, neither book brought him great wealth. He did use his royalties to visit his family: "I went there when Cameroon was still a French mandate. I returned in 1954, then in 1958, and finally in 1959, a few months before independence. But I never returned after that, for a very simple reason: In my country, opponents are not tolerated." (I: 151) He was informed by his friends that he should not return, because he had angered whites in the colonial power structure of which the religious missions were an integral part. His book was banned in Cameroon.

Like other African intellectuals who arrived in France and enjoyed some critical success, Beti became discontented with the French model of assimilation. Not only did he come to realize that engaged anti-colonial writers were seen as dogs that bit the hand that fed them, but he also arrived at a sophisticated understanding of the colonial cave in which francophone Africans lived. His writing in the 1950s directly challenged the core of French academic work on Africa, which was heavily influenced by ethnology. It was common in those days for educated French elites working on Africa to adopt an ethnological perspective. The work of Claude Lévi-Strauss is but one example of the kind of influence that ethnology had on the intellectual climate of the day. In particular Beti became conscious of the idea of African "tradition" as a discourse of domination. "The idea that, in daily life, the main preoccupation of Africans is to protect their traditions is a white myth of the colonizer. Africans move in a very rich reality where tradition is but one given among numerous others. Like all other peoples they relate in changing ways to their traditions by adapting to necessities that are endlessly renewed." (I: 47) He pointed to the conversion of Africans from animist traditions to Christianity. "What is *Le Pauvre Christ de Bomba* but the process of substitution of one religious tradition with another?" (I: 48)

He confronted the ethnological desire to study and capture African traditions, even to protect them, as an insidious form of cultural imperialism. He challenged "ethnologism," a false approach "which is not ethnology," as giving African traditions "an image outside of space and time" (I: 48): "The pretention to understand through African traditions not a moment of their being, but their essence itself, their intimate and definitive truth, the temptation almost always irresistible to consecrate certain attitudes, usages and habits that are in reality only functional, subject to wearing down, and of very relative significance." (I: 49) Colonial ethnologists created fixed abstractions that valorized strangeness and accumulated ultra-conservative traditions stored in their notebooks like freezers. "The consequence is that tradition, instead of being studied in relation with the life of its people and to improve it, is petrified into an autonomous entity, a sort of self-nourishing religion. It has become dogma." (I: 50) "It is wrong to always arrange beliefs, institutions, rites, customs, legends, and social hierarchies like a hunter hanging up trophies on the wall. From an historical perspective, the landscape of the ethnologist looks flat and sterile, while in reality, some traditions are in decline, others at their peak, and still others just emerging." (I: 52) "Like all other societies, African society if it is going to survive is condemned to transform, and renounce at least some of its traditions, or change them." (I: 59)

Not only did such writing make him unpopular among the specialized class of French ethnologists, but it also brought him into direct confrontation with the dominant current in francophone African writing: that is, *négritude*: "While a certain negritude takes honor in singing the praises of all African tradition, I think it is evident to the reader of my novels that there are good and bad traditions, and at any rate it is not enough that we have inherited a tradition from our ancestors for us automatically to venerate it." (I: 53) Beti vehemently lashed out at writers such as Léopold Senghor, whose version of negritude he saw as collaboration with white supremacy: "Negritude is neocolonial ideology." (I: 56) He often cited a famous aphorism of Senghor: "Reason is Hellenistic; intuition is Negro." Beti believed that this idea, gladly propagated by whites, portrayed blacks as irrational, incapable of reason, motivated only by primitive passions. Such ideas had a danger of becoming interiorized. Under the pretence of liberating Africans, Senghor's ideology of negritude kept them in chains.

Having offended the French bourgeois readers of fiction, the French ethnologists who dominated African studies, and the French literary critics who were singing the praises of negritude, it was clear that the literary career of Mongo Beti had come to an end, at least from a commercial point of view. Despite growing interest in his novels overseas, especially in the United States, none of his novels was reprinted in France. His stubborn refusal to change his rebellious stance led to him being blacklisted among the tiny group of influential critics who make or break literary careers. "There are notably a group of people in France who constitute the techno-structure of cooperation and *francophonie* who want me to make grand proclamations in favor of *francophonie*. They want to 'Senghorize' me because they think that an African intellectual should necessarily be a Senghor." (II: 21) "It is above all education, mass media, political manipulation, all of the influences of latent imperialism in our traditions" that this French intellectual elite sought to manipulate: "Because these are what constitute the essence of our identity." (I: 61)

The fundamental idea underlying all of Mongo Beti's work was that *writing is political*. "Whether he likes it or not, the gentleman or lady who writes performs a political act. Whether he is silent, or he speaks; in either case he takes a position." (I: 159) "If there is a chance for black literature in *la francophonie* to discourage the persecution by power, it is, in my humble opinion, the degree to which it refuses to mask the political dimension of its essence. At any rate, that is the only chance for African writers in *la francophonie* to get out of the diabolical traps of marginalization and recuperation." (I: 107) He came to have a larger grasp of the structure of cultural imperialism that imprisoned African writers in the French speaking world: "A people in general who are weak and impoverished, decolonized more or less in reality, among whom appear writers obliged to create in the language of their former colonizer, having become their protector but jealously guarding all the privileges of domination," (I: 99) were hard pressed to escape this postcolonial cave. "How can the preponderant institutions of *la francophonie*, that is to say, of the French academies, French publications, French universities, and the great French publishing houses, all tied to the history of France and thus to the enslavement and oppression of blacks, forbid themselves from persecuting a black writer who defies them?" (I: 102)

He drew parallels between the exploitation of African writings and other raw materials. "For a long time a flagrant fundamental antimony has existed between the engaged African writer, whose combat is nothing less than the liberation of his race, and the Western capitalist publisher, that speculator in printed paper obsessed with maximizing profits, his only law." (I: 229) "On the one side, the francophone Africans create a veritable raw material. But on the other side, the critics, commentators, and non-African professors of *la francophonie* reserve for themselves the exorbitant privilege of transforming this production into a product of consumption, becoming as unrecognizable as a tablet of chocolate produced in Hamburg or Marseilles to the African peasant who cultivates its cocoa." (I: 105) "The French university, through its technical assistants, cooperation workers, intimate counselors to presidents and ministers, accommodate themselves to the monstrous censorship which, in front of the eyes of everybody, is killing all creative initiative by francophone Africans." (I: 201) "Twenty years after independence, the most decisive diplomas are still handed out by French professors in French universities that are, not without reason, subordinate to French interests in Africa." (I: 203) French African writing was screened through neocolonial filters.

"In French Africa, there are two creative currents in literature: one on the right, and the other on the left. On the right, there is Senghor, Camara Laye, and also Ousmane Socé, the author of *Karim*. Those guys have always been well supported by French critics. On the other side there is a leftist current, other writers like Abdoulaye Sadji who have never been republished." (I: 160) "Me, I'm a man of the left. I'm a partisan for the evolution of things. I'm a partisan of progress, not just technological progress, but psychological progress, the progress of a group, its faculty to adapt to new situations. Everything that permits blacks to become the masters of their own affairs is progressive." (I: 177) "Only Africans on the left can really lead the way to emancipation of the people. Their real friends in the West and elsewhere in the outside world are parties on the left. Of this I have no doubt." (I: 178) "Only a certain kind of man can really change things, make a revolution in our house; men who are shaped by a leftist ideology, by socialism." (I: 179)

He excoriated editorialists in major newspapers who praised the soft authoritarian leaders in the former French colonies, the Goncourt prize-winning writers who sold African misery in the supermarkets, the documentary film-makers who produced lying dithyrambs. African writers were not spared in his critique. "The elite, in the largest sense of the term, in particular its *clercs*, has always been prompt in betraying their true mission for a little gold." (II: 44) At a conference in Germany he gathered numerous anecdotes illustrating this treason of the intellectuals: Lenrie Peters, the Gambian poet, demanding incredulously why he had not been paid (I: 224); Pierre-Makambo Bamboté, the writer, who said of politics that "nothing is worth sacrificing the creative needs of the artist" (I: 219); or Ahmadou Kourouma defending the Ahidjo regime: "I didn't say that I defended censorship. I only said that censorship exists in all countries of the world, more or less, and that it is not enough to judge a regime's politics." (I: 240) Needless to say, he did not make a lot of friends in exile.

Worst of all were those who claimed that Beti was not an authentic African writer because he did not live in Africa. Jean Dodo approached Beti at this conference: "Your exile has been extremely long and is regretted in your country. Would you like me to arrange your reconciliation with President Ahidjo?" (I: 242) As if being granted a visa to meet with a dictator was worth selling out all of his ideals. Such well-intentioned people did not seem to understand the danger of Beti returning to Cameroon, nor the injustice of making such open compromise the standard for evaluating his status as an authentic African writer.

At one point Beti was invited to present at the annual convention of the African Studies Association in Florida, but was unable to receive funding from that organization because he did not live in Africa. It preferred to sponsor *real* African writers. "From a sentimental point of view, I would love to be in Africa," he explained, "but from a pragmatic point of view, in terms of effectiveness, a francophone African writer has a real interest in staying here in France, because he can't live at home if he opposes the regime in power." (I: 155) "Since francophone Africa is still governed from here, it's here that you get the best information. If you want to know what's really going on in Cameroon, it's not in Cameroon that you will find out. You need to be in Paris. If you want to know what Ahidjo is going to do next year, you won't find out in Cameroon. It's in France that you are best informed." (I: 156)

Beti had passed his teacher's certificate in 1959, defended his doctoral thesis in 1961, and got married to a colleague, Odile, in 1963. In 1966 he passed the aggregation in classics, one of the most difficult competitive exams in the entire French university system, earning the prestigious title of "Professeur Aggregé." His recompense was to be named a high school Latin teacher in Rouen, at the Lycée Corneille, where he worked for a miserable salary under his legal name, Alexandre Biyidi-Awala. Despite having published some critically acclaimed novels, and reaching the highest level of academic achievement in France, there was a price to pay for his dissent. He had a wife and three children to take care of now. Gone were the dreams of becoming a famous writer. Blacklisted by the Paris literary community, he published nothing between 1966 and 1972, marginalized into a kind of professional oblivion as a schoolteacher in Normandy. "The high school, even the school board in the city of Rouen, painfully accommodated the presence of a black professor in the most prestigious lycée in Normandy. But I made this situation even worse for myself by acquiring a reputation as a controversial protestor against French policy in Africa. It didn't take long for me to be known as a Marxist agitator, even pro-Soviet." (I: 193)

He was not a communist, he later explained, but a Marxist. "I don't believe in tribal conflicts, but rather in class conflicts between the rich and poor. Stalinism is an abomination; Maoism, an unqualified heresy. Those were just erroneous interpretations of Marxism. Like the Inquisition to Christianity, it's not because cynical ignorant people dirtied the name of Marxism that this ideology should be condemned." (I: 17–18)

MAIN BASSE SUR LE CAMEROUN

The turning point in Mongo Beti's career came in 1972 with his publication of a political tract entitled *Main basse sur le Cameroun*, written after the Ahidjo regime arrested, secretly imprisoned, and tried the revolutionary Enrest Ouandie and a Catholic bishop, Albert Ndongo, for attempted coup d'état. No visitors were allowed, no lawyers were granted, and their location was kept a secret. "For me, as for the large majority of people, the former man was a hero, and his arrest and execution was to be expected. On the other hand, the persecution of a Roman Catholic bishop intrigued me." (I: 67) Ouandie was executed, while Ndongo was left wallowing in the filth of a concentration camp. "Until then I had given French neocolonialism in Cameroon a sort of current-affairs interpretation. It was in discovering, little by little, who Mgr. Ndongmo was, and what his projects were, that the Cameroon of Ahmadou Ahidjo appeared to me, for the first time, as a veritable bantustan where all the profits went to white businesses, and all the tears, all the wounds went to the negroes." (I: 71) Until then he had criticized the governments of Cameroon and France independently. But it was only in writing this book that he began to comprehend exactly how the former had collaborated with the latter in a permanent neocolonial system, "*Françafrique.*"

Ahmadou Ahidjo was the first president of Cameroon (1960–82) who still today enjoys a carefully polished image in the French-speaking world. Very little criticism of his more than two decades of single-party rule passed through the conservative filters of the francophone technostructure. You probably have a vague, positive image that pops into your head when you hear his name, if you have any image at all. The less you know about Ahidjo, or even Cameroon, the better this image will be. Perhaps you know him as the "founding father" of Cameroon who led his country into independence. Perhaps when looking at the surrounding states you imagine his era as an island of "peaceful civilian rule" in an otherwise violent region of coups d'état and military regimes: Ahidjo was no Babangida (Nigeria), no Habré (Chad), no Bokassa (CAR), no Sassou-Nguesso (Congo), no Nguema (Equatorial Guinea). Perhaps you have heard that he provided "national unity" in a country divided into 300 ethnic groups, with an additional cleavage between its anglophone and francophone communities. Perhaps you have even heard that he voluntarily stepped down and handed his office to his successor, President Paul Biya (1982–present), in a "peaceful transition of power."

But these are shadows of a reality that is very different when you read *Main basse sur le Cameroun*. Ahidjo faced several challenges. These included the growing guerrilla resistance to his rule by the Union des Populations du Cameroun (UPC), the true independence party that had first called for an end to French rule, defended the need for rapid improvement in the living standards of the people, rejected the incorporation of Cameroun into the French Union, and advocated the immediate reunification of French and British Cameroons as a precondition for independence. By crushing the UPC rebellion in 1958, with the help of French troops, Ahidjo was in reality a puppet of French interests, a collaborator with the larger neocolonial project of Charles de Gaulle and Jacques Foccart. He brought Cameroon, a former German colony divided into a French and British mandate, into the neocolonial network of French Africa. He consolidated all power into his hands by creating a one-party structure under the Cameroon National Union (CNU) and serving as its permanent chairman. He was also the architect of a veritable police state, which censored information and silenced critics with a security apparatus armed and trained by the French. For Ahidjo, the ends of unity justified the means of violence. For Beti, Ahidjo was a dictator.

Re-elected without an opponent (non-democratically), Ahidjo effectively silenced all opposition with the execution of Ernest Ouandie in 1972. Ouandie was a revolutionary leader of the UPC guerrilla forces in the anglophone western province (former British

Cameroons), who had passed into armed resistance when his party was officially banned. "In 1970, Ahidjo had been exterminating the left in Cameroon," Beti explained, "to the grand pleasure of [French oil company] Elf-ERAP and Pechiney, who were exploiting the energy resources of the country with the kind of profits that one wouldn't find in a fairy tale." (I: 277)

Petroleum deposits had been known to exist in Cameroon as early as the 1950s, but it was not until 1972 that permits were granted to foreign oil companies coming from France, the Netherlands, Norway, Canada, and the United States. Ahidjo maintained a cloud of secrecy over oil production and export. Neither production nor export figures were published. The price paid for Cameroon's oil also remained a state secret. Most disturbingly, the Ahidjo government did not disclose the final destination of oil revenues, nor include them in the national budget. This tactic was known euphemistically as *le compte hors budget,* or the "extra-budgetary account." Ahidjo's justification for these official government policies—a lack of transparency that today would be called grand corruption—was that the economy should not be based entirely on its petroleum revenues. According to Ahidjo, agriculture was the mainstay of the Cameroonian economy because "before petroleum there was agriculture and after petroleum there will be agriculture." (DeLancey 2000: 220) Of course nothing stopped petroleum revenues from quickly surpassing all other sources of revenues. Nor did his keeping oil revenues a state secret do much to channel them into poverty alleviation or economic development.

It was while Beti was researching this new oil economy of his native country that he started to understand how French colonialism had transformed into French neocolonialism; that French domination was not a passing phase, but a structural reality; that France was "helping itself" (*main basse*) to the resources of Cameroon, as the title of his book put it. Moreover he came to understand that unless people understood what was happening, a new and greater evil than the authoritarian regime would emerge in Cameroon. In a letter addressed to French president Valéry Giscard d'Estaing, Beti accused the French of using their puppet Ahidjo to keep his country poor and undeveloped: "When Elf-Aquitaine succeeds in exploiting the oil of Cameroon without paying compensation to the indigenous people, and during an era when this black gold is turning Arab countries into Eldorado, we Africans are seized by a most righteous indignation." (I: 258)

Beti spent a year and a half researching and writing his book in secret. "A few weeks before the book was scheduled to come out, that is, around mid-May 1972, an inspector from the *Renseignements Généraux* who had taken the habit of visiting my apartment near Rouen in the afternoons when I was absent, asked my wife about my current literary activities, my nationality, and my profession." (I: 75) Judging by his attitude, their hope was that this psychological pressure would suffice, would doubtlessly intimidate me, and would thereby lead me to renounce the publication of my book." (I: 76) "One afternoon, after a chapter of the book had been published in the *Partisan* these gentlemen came to my school and, with the strict minimum of courtesy, ordered me to follow them to an empty classroom. Once there, they told me that they had been sent by the police chief to verify my papers." (ibid.) "The few documents that I ceremoniously took out of my wallet were clearly not enough: Was I, or was I not, a foreigner in France? In other words, could my book be legally prohibited and seized when it came out?" (I: 76–7)

He told them he had a French passport at home. That evening two policemen came to his apartment and demanded to see his alleged document; they were consternated to find that he did in fact posses one. "Born in Cameroon, a country then directly administered by France, I had been issued, as a French subject, a French passport when I had arrived in France in 1951. But it was true that I had been deprived of a national identity card." His whole adult life had been spent in France, and thus he said that he had considered himself French by default. But his passport had expired, so the police told him he would have to be "naturalized."

Main basse sur le Cameroun was the object of censorship by a government prohibition decree that appeared in the *Journal Officiel*. This prohibition was authorized by an old 1939 law against German propaganda, which stipulated that foreign works could be censored in the name of national security. In fact, the law specified "works coming from foreign countries," which provided a basis for his subsequent legal challenge. But the book was seized from the publishers one week after its release. This blatant act of censorship was later the subject of a documentary called *Contre-Censure*, produced by Canadian journalists with the support of the Canadian section of Amnesty International (but suppressed by the French section, from which he was excommunicated in 1976). Censorship turned him into a celebrated African dissident.

For the next four years he waged a legal battle to have his book uncensored. He underwent a series of Byzantine administrative procedures designed to have him extradited to Cameroon. His request for a new passport, for example, authorized by a French judge (the normal procedure), had for some reasons not been signed. When he went to the police prefecture to have this corrected, he was told to go to "his" embassy to make a new application. This was a trap: if he went to the consulate he would place himself in the hands of the Cameroonian authorities, the very ones who had pushed for his censorship, and who wanted to have him arrested and imprisoned. He refused. In the end, his legal challenge prevailed. *Main basse sur le Cameroun* was re-released in 1976. Mongo Beti became the living symbol of opposition to both the Ahidjo regime (perhaps its only internationally recognized critic) and the gadfly of French neocolonialism in Africa (perhaps the only one with a French audience).

"Paris and Africa today is Washington and Latin America. Paris favors, not only in Cameroon, but in Africa, the putrid over the upright, gangsterism over probity, the past over the future, the sclerotic tradition over the inventiveness of youth, cynicism over hope, and reactionary immobility over African revolution." (I: 333) In 1980 he published a famous open letter to President Giscard. "While your predecessors during their 14 years only sent troops to Africa three times (Cameroon in 1960, Gabon in 1964, and Chad in 1984) you have beaten all the records in only five years as president, having sent your parachutists twice to Zaire to help your friend Mobutu Sese Soko, once to Chad under Félix Malloum, once in Mauritania to help a junta of colonels, and now once more in Central Africa to install David Dacko, your new protégé." (I: 251) This kind of invective was classic Mongo Beti—he understood his responsibility to speak truth to power. "Both the silence of Africans, and French military intervention, are two faces of the same monster." (I: 252) In his African version of Zola's *J'accuse*, Beti criticized Giscard for accepting $10 million worth of diamonds from Bokassa. "Imagine all the schools," he asked Giscard, "imagine all the maternity wards," and "all the kilometers of paved roads" this money could have provided: "You didn't hesitate bending over to pick up some spare change out of the putrid gutter that is Bokassa's Central Africa. What strange depravity could inspire a rich man to plunder and deprive a poor beggar?" (I: 261) The diamonds scandal was a factor in Giscard losing his bid for re-election.

THE RETURN TO CAMEROON

The 1981 election of the Socialist President François Mitterrand promised to change French African policy. But after their initial euphoria, leftists such as Mongo Beti became disappointed. In another open letter published in 1991, Beti criticized Mitterrand for ten years of "imitating your right-wing predecessors." (II: 69) He was particularly disappointed by the continued support of Mitterrand for the new regime in Cameroon. "In 1982, obviously under your patronage, Paul Biya replaced the previous dictator, Ahmadou Ahidjo." Using the pretext of liberalization, he wrote, Biya had increased the "corruption, embezzlement, waste, and tribalization of the civil service, police despotism, repression, and the collapse of public finances." (II: 70)

The only thing keeping even a semblance of national unity in Cameroon was external support: "The Biya system could not survive without being subsidized by France." (II: 73) In this open letter Beti demanded that Mitterrand stop supporting the regime with the pretense that it had been democratically elected. "Paul Biya elected? He was never elected. On the contrary, he came to power through elections in a one-party system. Words must be given their proper meaning. What is an election? The root of this word comes from Latin, and means 'to choose,' that is, I have several objects before me, and among them I choose one." (II: 76–7)

In 1991 Cameroon entered into an economic crisis, caused in part by the decline in oil revenues, and in part by massive debt (itself a result of institutional corruption and foreign exploitation). This economic crisis triggered a political crisis (Takougang 1997). Biya was forced to open up the political system to multiparty politics in 1989, which unleashed a wave of protests and anti-system activism by mid-1991 that led to a state of emergency and most of the territory being placed under military rule. Biya was an ethnic Beti who pretended that this fact had privileged the center–south region near Yaoundé, but this tribalism did not endear him to Mongo Beti, who remarked that ethnic privileges granted under Biya had only furthered the division of the people in the anglophone West, coastal Douala, and Muslim North. "The country is on the verge of falling apart." (II: 72) It was in this climate in 1991 that Beti, tired of dealing with the surveillance and intimidation that he suffered in France, made the decision to return to Cameroon. "After *Main basse sur le Cameroun*, that is, for 20 years, each time I published a book, I warned my wife that we were going to have problems.

Almost always, during that period, the police would literally siege our apartment." (I: 370) He was followed through train stations on his work commute. "I was tracked like a criminal, because I was a black writer who paid no allegiance." (I: 399) "For a long time I asked myself, why can't I write freely and happily, like other writers of my adopted country? Now I know the answer. It was so simple, like all great truths. *Eh bien*, it's because I am an African writer, and my true mission is to prepare the way for African writers who will follow, so that they can write freely and happily." (I: 401) Following his legal battle against the censors, he wrote and published several important works of fiction, including *Remember Ruben* (1974), the title of which invoked the famous independence fighter Ruben Um Nyobé, assassinated by the French in 1958; then *La Ruine presque cocasse d'un polichinelle* (1979) about prostitution in Cameroon; and a pair of social realist novels inspired by Zola, *Les deux mères de Guillaume Ismaël Dzewatama* (1983) and *La revanche de Guillaume Ismaël Dzewatama* (1984). After three decades of publishing novels in France, with secondary publishing houses that provided him a limited commercial success, he realized that he was never going to win a Goncourt. "I am a writer, but I will never be one of theirs." (I: 399)

After 32 years in exile he returned to Cameroon for a week of conferences in 1991, upon the invitation of members of civil society who were preparing for a National Conference. He had a hard time getting a visa to enter the country from the Cameroonian embassy in Paris, and he was almost arrested at the airport upon his arrival. "I was shocked by the welcome of the police at the airport. They submitted me to a humiliating search, and sequestered me, which showed that Cameroon was, as I had been told, a police state." (II: 53–4) But most of all he wanted to see his mother. "One of my cousins, a man of the entourage of the dictator Ahidjo, had regularly visited my mother to tell her lies. He said he had visited me and spoke to me about her misery, sickness and despair. He told her that I had forgotten her completely." (I: 395) "I knew my mother was in a state of misery and I had tried many times to send her money, but it was always confiscated by the political police. If one of my relatives was suspected of writing to me, he was arrested, detained, interrogated, terrorized. One of my 10-year-old nephews, whose dead father was my younger brother, experienced this." (ibid.)

In February of 1991, "I found my mother, an old octogenarian, grimacing, in the place of the young woman with an easy smile I had left behind 32 years before. Having worked hard all her life, she

suffered from acute arthritis, and she didn't see anything anymore."
(I: 397) "I went back twice in that year to take care of her. Then
in the second week of December, she was hospitalized. Her health
rapidly deteriorated. My mother died ten months after I had found
her again." There was no clinic in the village, and she was poor,
so she could not afford treatment. "This suffering is a lesson for
young Africans who come to hear my story, and who will certainly
not miss meditating upon it before entering the field of literature,
an activity which will require they wait a long time before knowing
even a little happiness." (I: 138)

Members of the ruling party of course accused him of coming
back to Cameroon to run for president. "It's completely stupid,"
he told a reporter. "I have a social status that entirely satisfies me:
writer and educator. That is my vocation. The proof is that it is
these two professions which I have practiced for 30 years. I am not
fighting for politicians, but for principles like freedom of speech, of
the press, for fair and free elections; I am fighting so that France no
longer confiscates Cameroon's oil resources, that its management
is transparent, and that oil revenues are no longer the privilege of
rulers alone. That is what I'm fighting for." (II: 76–7)

In 1993 he retired. Receiving a teacher's pension that allowed him
to live, he decided to move back to his native village in Akometan.
The risks were great, but down he went. "What struck me upon my
return was the destitution of Beti countryside. Nothing had been
brought by thirty-five years of independence. Akometain had three-
thousand-five-hundred inhabitants, of which only three-hundred
kids were in school. The people still drank ground water, had no
medical center in the village, and did not even have a general store."
(II: 168) "My work was to bring some modernity to this medieval
village, starting with drinking water. I drilled some wells, and built
a second school for my village. I also intended to create a medical
center to prevent malaria. I needed money to do all of that, but I
couldn't get a loan from the regime, and the only resource available
to villagers was their ancestral forest." (II: 169) So he opened a small
artisanal woodcutting operation. The authorities quickly seized
the wood, claiming that the forests belonged to the state. Biya was
selling vast forestry concessions to foreigners, but he would not
permit the people to use their only remaining local resource. "They
took our oil to reimburse their debts, and now they are taking our
wood to reimburse their debts to the international lenders." (II: 173)

He also opened a bookstore in Yaoundé called *La Librarie des
Peuples noirs*. "I was disheartened to discover the scandalous yet

organized scarcity of books in Cameroon. What shocked me was that there were no books; that in a classroom with over a hundred students, which is common here, you won't find a math textbook, to understand the dire situation. It is impossible to imagine that the French authorities don't know about this." (II: 276) "It doesn't shock me that there are no political books, given the political system we're talking about. But there is a veritable sabotage of the circulation of any and all books. Cameroon is the only African country which still imposes a customs duty on books entering it, doubling the price of a French book in a country where the public is forty-times poorer." (ibid.) "In Yaoundé, for example, the capital of Cameroon, a community of one-million-five-hundred-thousand inhabitants, there is not a single bookstore accessible to the people. It's barely believable, here at the beginning of the third millennium. Here was a community of one-million-five-hundred-thousand people without a library, never producing a researcher, or a poet, or a novelist." (II: 282)

Having descended back down into the cave, for the next ten years he played an active role in the political and cultural life of Cameroon, frequently giving interviews to the press and audiovisual media. He continually denounced the neocolonialism of France, publishing *La France contre Afrique* (1993) which exposed the consecutive collaborationist regimes of Ahidjo and Biya with the French oilmen and the foresters. His writings on oil were nourished during this period by scandalous revelations in the French press surrounding the "Elf Affair." "I know personally through a reliable source that the French government [then under the French socialist Lionel Jospin] has resigned itself to accepting the latest re-election of Paul Biya because of lobbying pressure by Elf-Aquitaine, which reigns over our political destiny. Lionel Jospin himself has said that he is powerless when faced with the omnipotence of this oil company. We need to re-organize our strategy of national liberation to conform to this reality. What good is it to accuse an abstract France when representatives admit they have no power over their oil firms? One day we will have to take on, *hic et nunc*, in one fashion or another, all the powerful actors, whether French or otherwise, that support Paul Biya, and which are thus instruments of our oppression. That includes Elf-Aquitaine." (III: 239)

"The Elf-Aquitaine scandal illustrates the cynicism of a foreign policy and egocentrism of a great power being haunted by its hypocrisy. Not only does the exploitation of oil bring nothing to Africans, but it reinforces their oppression, by furnishing ever increasing means of repression to dictators. Moreover, it is the source of super-profits that give this business the appearance of the slave trade. Why should we allow Elf to reinforce in our country a mafia of assassins, bearded mercenaries, and other con-men who want to construct a pipeline from Chad through Cameroon?" (III: 240–1)

But Beti mainly denounced the failure of public services in Cameroon, the postmen who stole their parcels, the banks that stole money deposited with them, the corruption of schools that demanded bribes from parents yet provided no books, the predatory policemen who used violence and intimidation, the ministers who embezzled their budgets for personal enrichment while public infrastructure fell into decay, the complete absence of rigor under Biya's regime. Beti did what he could to transform consciousness before he died on the night of October 7, 2001, in Yaoundé, at 62 years of age. Take a look at his bibliography, before turning to the next chapter. For that is the legacy of an engaged intellectual, or, to use his preferred term, a "dissident."

During the last years he wrote an essay about his personal reflections on dissidence, in which he addressed its apparent failure in Cameroon. "I was talking the other day with a sympathetic young Cameroonian who told me, "*Monsieur*, you should be discouraged, because you have failed. Since you have started your battle, what results have you achieved? I wasn't even born when they were talking about your struggle. Yet nothing has changed!" My interlocutor was right, but only partly so. It is true that we are still under the jackboot of a dictator whose power is entirely based on the violence of the police, so that innocent citizens can be beaten with impunity in public. It is true that, unlike other peoples in the world, we Cameroonians still cannot freely elect our leaders. It is true that we are more dependent on foreigners than ever, from all points of view. Et cetera, et cetera. But it is also true that we have been for some time now experiencing a political turbulence which, paradoxically, is symbolic, or if you prefer symptomatic, of our emergence into modernity. Dissidence is one of those symptoms." (ibid.)

In "Quelques réflexions personnelles sur la dissidence," he attempted to prepare the way for fellow prisoners to follow his footsteps out of the cave. He outlined the responsibility of the Cameroonian intellectuals: "Dissidence is a phenomenon that concerns above all the social category called the intelligentsia: writers, like Rousseau, poets, dramatists, painters, film makers, researchers, that is, in our day, university professors. Why the intelligentsia? Because they are the national category whose conscience is most demanding, most sensitive, and thus, most tormented. It is no accident that, in all the situations of oppression and injustice, it is the intellectual whose protests are first heard." (II: 221) "The dissident is someone who voluntarily at a given moment separates himself because he doubts the values upon which his group is based." (ibid.) He is reacting to unhappy circumstances; first of all, to his own unhappy circumstances. From the start he comes to reject society's values of success. "Dissidence is not born of ideological speculation, but life experience. That is what explains its force and, even more, its faculty to create new values in a society, like ours, where vulgar success, that is to say, money, has sterilized everything. Despite its apparent passivity, what is proper to dissidence is in effect its ability *to create a new ethos.* Our Cameroonian dissident intelligentsia, as anyone can observe, drafts an ideal that contradicts the practices forged by our political leaders, teaching the little people that they can live and act otherwise." (II: 223–4)

"Dissidence is often followed by a rising movement. Not only does it increase in quantity (that is to say, there are more and more dissidents), but the atmosphere created by the presence of these dissidents makes itself increasingly heavy, as if the nation is heading towards serious events, for example, a revolution." (II: 224) He cites Soviet dissidents such as Mayakovski, Pasternak, and Solzhenitsyn, whose dissident writings about gulags and psychiatric hospitals "gave a kind of death blow to the image of the system." (II: 225) "This inundation is the most striking characteristic in Cameroon at this moment. One might say that dissidence has become the natural manner of being for the intelligentsia. In fact, what is an intellectual, a writer, a filmmaker, a Cameroonian artist who is not in a situation of dissidence? It is precisely this explosive phenomenon, more than all of the other aspects, which might permit us to give a sense to dissidence." (ibid.) "Perhaps we are not yet at that advanced stage here in Cameroon. Would it be desirable anyway? At least to see in it a symbol of the extreme failure of the political system." (II: 226) "To conclude, dissidence appears as a specific response of the intelligentsia, the most enlightened class of a nation, to political constraints, which in our country are the constraints of a neocolonial dictatorship. The more that dissidence assumes, as it is among us now, a mass character, the more patent will be the damnation of the political system which generates these constraints." (II: 227)

"Far from being a separate part, the intelligentsia is in fact an integral part of the nation. That is to say, dissidence is but the visible part of an iceberg, which is nascent popular resistance. Soviet dissidence, when it became massive, reflected the immense weariness of the Soviet people. They could simply take no more. Here, like there, dissidence signifies that the state and the people are in a situation of hostility, that is, a cold war against nature. At the same time, paradoxically, they are burning for reconciliation: For the state and the people must live in harmony. That is what is called democracy. But for this reconciliation to take place, the system which has perverted our state into a wolf that consumes those it was supposed to protect, first must disappear. This is how reconciliation between a regenerated state and the Cameroonian people will happen, through democracy put into service of liberty." (ibid.)

Having written what could be considered an obituary of Mongo Beti, the question remains if we should praise him or to bury him. How successful was he? Measuring the success of an intellectual dissident requires one to adopt a very long-term perspective. (Solzhenitsyn is being taught today in Russian schools. Is that a measure of his success?) If we fix our regard on the sad story of the press in Cameroon, we observe that out of 1,300 private newspapers that came into existence following the 1990 reforms of its media law, only 60 still exist today. (Atenga 2007: 23) In this banana republic with its simulacra of pluralist elections, the free press is exposed to arrests, beatings, torture, imprisonment, seizures, censorship, warnings, menaces, and a judicial system that makes criticism of the president a criminal offense. A journalist can be condemned in court for the possession of administrative documents, and when he cannot provide such proof to support his allegations, the punishment is even greater. The repression of the free press in Cameroon is one of the principal tools allowing the longevity of despotic personal rule.

Nevertheless, looking at Cameroon since Beti's return in the early 1990s, for the first time in the political life of this country, the cataclysm affecting ordinary citizens has been made public. The press has taken advantage of its margin of freedom, and opposition newspapers have survived. Editorial language is no longer antiseptic or permanently euphemized, and a political discourse has emerged that marks a rupture with official dogma. Writings of the press have allowed for a new linguistic ethic to acquire legitimacy. Journalism has gained a creative power to invoke a plurality of perceptions. The media rupture with state dogma is an ideological opposition to power. Beti influenced the editorial style of Cameroonian dissidents, and young writers now understand that their regime is sensitive to the scathing words that strip it naked.

7
Political Parties and Elections

ELECTORAL DEMOCRACY IN AFRICA

Some people believe there is no such thing as the oil curse. They point to countries such as Norway, which are highly dependent on their oil revenues, but do not suffer from any "resource conflict," nor symptoms of a "rentier state," nor an economic "paradox of plenty." Or they point to other oil-producers such as Scotland and Alaska, which seem to suggest that the real problem for the African oil states is not their oil, but their lack of democratic governance. If these countries democratized, they argue, then oil could become a blessing and not a curse. This chapter will concern itself with the potentials of democratization, and in particular with the possibility that democratic elections and opposition parties can fight political, economic and conflict syndromes of the oil curse. Can Africa replicate the positive experience of Norway? Is multiparty democracy feasible in African oil states? And if so, can political parties transform their structural economic conditions?

First we must define what is meant by the word "democracy." For many years during the Cold War a strict procedural definition of the term was accepted. Joseph Schumpeter's minimalist definition of fair and free elections seemed to be a valid distinction between democracy and authoritarian systems: "The democratic method is that institutional arrangement for arriving at political decisions in which individuals acquire the power to decide by means of a competitive struggle for the people's vote." (Schumpeter 1947: 269)

But during the "third wave" of democratization (Huntington 1991) when the number of new democracies rapidly increased, their elections did not always produce democratic outcomes. Not only were authoritarian regimes capable of staging unfair elections to please foreign donors and critics, but even in new "transitional democracies" the weakness of many other institutions and practices—especially civil liberties—raised doubts about thin procedural definitions of "electoral democracy."

Democratic theorists such as Robert Dahl had long argued that a more complete procedural definition was required; he coined the term "polyarchy" to capture the idea. (Dahl 1971) Of course, he argued, democratic practice requires officials elected in free and fair elections by universal suffrage. But it also requires other constitutional guarantees that protect the freedom of expression, alternative sources of information coming from a free press, and the associational autonomy of actors in civil society. (Dahl 1989) So Dahl distinguished two dimensions of democracy. The first was characterized by an enforceable set of rights and opportunities on which citizens may choose to act. The second was their actual participation in political life. To view democracy in light of only this second dimension is wrong, because "a country without the necessary rights and opportunities would as a consequence also lack the fundamental political institutions required for democracy." (Dahl 2000: 38) Such thick procedural definitions, with numerous criteria of civil liberties, measure what is called "liberal" democracy. Elected regimes lacking such civil liberties are called "electoral" or "illiberal" democracies. (Diamond 1999)

Scholars from Freedom House who rank-order the democratic freedom of countries by explicitly distinguishing between their political rights and their civil liberties (see Box 7.1) observed the rise of this new category—which they called "electoral democracy"— in the post-Cold War era: countries formally establish democratic electoral procedures, but do not protect constitutional liberties, nor limit the arbitrary exercise of executive power. In other words, many new transitional democracies seemed to have democratized only in terms of elections. "Illiberal democracy is a growth industry," declared one study of transitional democracies. "And to date few illiberal democracies have matured into liberal democracies." (Zakaria 1997: 24) This observation, like many others, is based on Freedom House data that rank political and civil liberties on a scale of 1 (completely free) to 7 (completely unfree), then averages these into an overall freedom score. Countries with an overall threshold score of 2.5 or less are "free." Those with averages of 3 to 5 are "partly free," and those with scores of 3.5 to 7 are "unfree".

Freedom House rankings are often used by people who do not understand how the numbers are arrived at. Without understanding the methodology, many people have a false impression that the numbers are picked out of a hat. While it is true that the rankings are generated by opinions of country specialists, these experts are not simply asked "how free is the country?" and then invited to pick a number between one and seven. The ratings are determined by a checklist of 25 questions, of which 10 address political rights, and 15 address civil liberties. Experts are asked to give a raw score between 0 (smallest degree of freedom) and 4 (highest degree). The political rights questions (total of 40 points) are grouped into three sub-categories (electoral process, political pluralism, and functioning of government), while the civil liberties questions (total of 60 points) are grouped into four sub-categories (freedom of expression and belief, associational and organizational rights, rule of law, and personal autonomy and individual rights). These 25 questions are broken down into numerous specific questions (*italicized* below in truncated form) operationalizing the concepts. In total, experts are asked over a hundred questions and provide evidence supporting their opinions, which are then examined by regional panels before assigning the final ranking.

Political Rights Questions

1. Is the head of government or other chief national authority elected through free and fair elections? *(Monitored by international organizations, without politically motivated delays, where voter registration non-discriminatory, using a secret ballot, with campaigns free from intimidation, a transparent vote-counting procedure, and each vote given equal weight.)*

2. Are the national legislative representatives elected through free and fair elections? *(Monitored by international organizations, without politically motivated delays, where voter registration non-discriminatory, using a secret ballot, with campaigns free from intimidation, a transparent vote-counting procedure, and each vote given equal weight.)*

3. Are the electoral laws and framework fair? *(No change in electoral laws immediately before the election, independent electoral commission with fair and balanced composition who conduct their work competently, with universal equal suffrage, fairly drawn electoral districts, without manipulation of electoral lists.)*

4. Do the people have the right to organize in different political parties or other competitive political groupings of their choice, and is the system open to the rise and fall of these competing parties or grouping? *(Encounter no undue legal obstacles of registration or membership, no discriminatory restrictions on meetings or rallies, with party leaders not intimidated, harassed, arrested, imprisoned, or subject to violent attacks.)*

5. Is there a significant opposition vote and a realistic possibility for the opposition to increase its support or gain power through elections? *(No selective restrictions on opposition parties, which hold real positions of authority, their leaders neither intimated, harassed, arrested, imprisoned, or subject to violent attacks.)*

6. Are the people's political choices free from domination by the military, foreign powers, totalitarian parties, religious hierarchies, economic oligarchies, or any other powerful group? *(No bribes of voters or leaders, nor intimidation, harassment, or attack of voters or leaders, where military does not hold preponderant control, nor foreign governments enjoy preponderant influence over policy, by presence of troops or economic sanctions.)*

7. Do cultural, ethnic, religious, or other minority groups have full political rights and electoral opportunities? *(Where parties address issues of specific concern to minorities, whose participation is not inhibited by laws or practices, but allowed to operate peacefully.)*

8. Do the freely elected head of government and national legislative representatives determine the policies of the government? *(Those elected are actually installed in office, without interference by non-elected state actors, including criminal gangs and military and foreign governments, in adopting and implementing policy.)*

9. Is the government free from pervasive corruption? *(Anticorruption laws implemented, bureaucratic regulations or registration controls not excessive, with independent auditing and investigating bodies, prosecuting allegations of corruption, given extensive media coverage, with legal protections for whistle-blowers, anti-corruption activists, and journalists.)*

10. Is the government accountable to the electorate between elections, and does it operate with openness and transparency? *(Civil society, interest groups, journalists, and citizens are able to influence policies, and have legal rights to obtain information, with a budget process subject to scrutiny, publication of government expenditures, transparent awarding of government contracts, and declaration of assets by officials open to the public and media.)*

Civil Rights Questions

1. Are there free and independent media and other forms of cultural expression? *(No censorship or self-censorship on*

politically sensitive issues, nor libel and security laws being used to punish critics, nor criminalization of insulting president or other government officials, nor political award of broadcast frequencies, nor journalists threatened, arrested, imprisoned, beaten, or killed)

2. Are religious institutions and communities free to practice their faith and express themselves in public and private? *(No registration impediments, nor members harassed, fined, arrested, or beaten, nor government appointment of religious leaders, government control of religious books or materials, construction bans on religious buildings, restrictions on religious education, nor mandatory religious education.)*

3. Is there academic freedom and is the educational system free of extensive political indoctrination? *(Teachers free to teach political matters, no government control of the curriculum for political purposes, no pressure on teachers or students to support officials or policies, no discouraging them from doing so.)*

4. Is there open and free private discussion? *(Discussions of a public nature without fear of harassment or arrest or people employed by government to engage in public surveillance, or reporting of anti-government conversations to the authorities.)*

5. Is there freedom of assembly, demonstration, and open public discussion? *(No banning or severe restrictions on peaceful protests, nor cumbersome legal requirement for permission to protest peacefully, nor intimidation, arrest, preventative detention, nor assault of protestors.)*

6. Is there freedom for nongovernmental organizations? *(No onerous registration requirements intended to prevent them functioning freely, nor cumbersome financing laws, nor pressure on donors, nor members intimidated, arrested, imprisoned, nor assaulted because of their work.)*

7. Are there free trade unions and peasant organizations or equivalents, and is there effective collective bargaining? Are there free professional and other private organizations? *(Allowed to exist, without government pressure on workers or peasants to join or to not join, without harassment, violence, or dismissal from their jobs if they disobey, with permission to engage in strikes without reprisals, and collective bargaining allowed.)*

8. Is there an independent judiciary? *(Free from interference from government or other private influences, where judges*

are appointed and dismissed fairly, their verdicts impartial, and compliance with judicial decisions by other governmental or powerful private actors.)

9. Does the rule of law prevail in civil and criminal matters? Are police under direct civilian control? *(Protection of defendants' rights, presumption of innocence, right to legal counsel, fair public trial, independent prosecutors, civilian control of law enforcement, free from organized crime, powerful commercial interests or other non-state actors.)*

10. Is there protection from political terror, unjustified imprisonment, exile, or torture, whether by groups that support or oppose the system? Is there freedom from war and insurgencies? *(No arbitrary arrests or detentions without warrants, no beatings of detainees, no use of excessive force to extract information or confessions, citizens can petition for violations of their rights, no private non-state actors engage in crimes, nor violence or terror due to civil conflict.)*

11. Do laws, policies, and practices guarantee equal treatment of various segments of the population? *(Targeted groups can exercise human rights with full equality before the law, perpetrators brought to justice, no persecution of discriminated groups, and gender equality.)*

12. Does the state control travel or choice of residence, employment, or institution of higher education? *(No political restrictions on foreign travel, no undue permission required for movement within territory, no government determination of type or place of employment, no bribes needed to obtain travel documents, change of residence, or employment, entry to higher education.)*

13. Do citizens have the right to own property and establish private businesses? Is private business activity unduly influenced by government officials, the security forces, political parties/ organizations, or organized crime? *(No undue interference with purchase or sale of land, adequate compensation for property expropriated under eminent domain laws, reasonable registration and licensing for private businesses, no bribes needed to obtain legal documents, nor extortion by organized crime.)*

14. Are there personal social freedoms, including gender equality, choice of marriage partners, and size of family? *(No violence against women including wife-beating and rape, or trafficking of women or children, no de jure or de facto gender discrimination, no laws for child marriage or dowry payments,*

no state-sponsored indoctrination nor undue infringement by private institutions including religious groups on marriage choices.)

15. Is there equality of opportunity and the absence of economic exploitation? *(No state price setting or production quotas, large state industries benefit general population and not the privileged few, no cartels, blacklists, or monopolistic practices by private interests, neither nepotism nor bribes for access to higher education, no discrimination against minorities, no unfair withholding of wages or forced work under unacceptably dangerous conditions, no adult slave labor or child labor.)*

This long battery of questions is one of the thickest procedural definitions of democracy. When applied to oil-rich African states, it is not surprising to find almost none pass the test. Some, such as Equatorial Guinea and Sudan, get the lowest possible ranking. Some, such as Nigeria and Gabon, receive a partly free ranking. But only the tiny republic of Sao Tomé e Príncipe qualifies as "free" in terms of its political rights and civil liberties (see Table 7.1).

Table 7.1   Liberal democracy in African oil states

	Political Rights (PR)		
Civil Liberties (CL)	"free"	"partly free"	"unfree"
"free"	São Tomé & Príncipe		
"partly free"			Nigeria, Gabon
"unfree"			Angola, Cameroon, Chad, Congo, Sudan, Equatorial Guinea,

Source: Freedom House, "Freedom in the World" (2009) www.freedomhouse.org

Of course, São Tomé is not yet producing any oil, so it is not yet an oil-rent-dependent state. But this is simply a matter of time. Its oil reserves are estimated to be in the billions of barrels, foreign oil companies have been paying big signature bonuses for offshore concessions, and everyone knows that sometime in the not-too-distant future São Tomé will be an important oil exporter. Therefore, because of its democratic governance, this country provides a unique case study to test the proposition whether "democracy" can resist the "oil curse" in Africa. Until the other states improve their performances, São Tomé is the only country in our sample where the Norwegian experience might possibly apply.

Also, from a comparative perspective the case study of São Tomé offers a contrast to the experience of Equatorial Guinea, for these two countries share many relevant characteristics. They are located in the same region of Africa, which gives them similar geographic location, climate, and regional history of slavery. Both have small populations. Both export cocoa. Both experienced colonial rule under Iberian fascist regimes (São Tomé under Salazar's Portugal, and Equatorial Guinea under Franco's Spain). In fact the number and relevance of similarities are so striking that, although history rarely provides us with the same rigorous control as experimental method, here we have an ideal comparative framework for "parallel demonstration" of two cases that are similar on many theoretically relevant characteristics, but differ in one important respect causally related to an outcome. The important difference here is the presence or absence of democracy. The important outcome is the presence or absence of the oil curse.

Box 7.1   How free is São Tomé?

"Freedom of expression is guaranteed and respected. While the state controls a local press agency and the only radio and television stations, no law forbids independent broadcasting. Opposition parties receive free airtime, and newsletters and pamphlets criticizing the government circulate freely. Residents have access to foreign broadcasters. Internet access is not restricted, though a lack of infrastructure limits penetration. Freedom of religion is respected within this predominantly Roman Catholic country. The government does not restrict academic freedom. Freedoms of assembly and association are respected. Citizens have the constitutional right to demonstrate with two days' advance notice to the government. Workers' rights to organize, strike, and bargain collectively are guaranteed and respected."

Source: Freedom House, *Freedom in the World 2010 www.freedomhouse.org*

MULTIPARTY DEMOCRACY IN SÃO TOMÉ E PRÍNCIPE

São Tomé e Príncipe consists of a pair of small Atlantic islands in the Gulf of Guinea. Unlike neighboring Bioko Island with

PRÍNCIPE ISLAND

SÃO TOMÉ
AND
PRÍNCIPE

*GULF OF
GUINEA*

SÃO TOMÉ ISLAND

São Tomé

Map 7.1 São Tomé e Príncipe

its indigenous population of Bubi living there before the age of European discovery, these two islands were uninhabited until the fifteenth century, when they were discovered by the Portuguese and transformed into the world's very first tropical plantation economy based on sugar and African slave labor. Since they were taken to these islands as individuals, and not as social groups, African slaves brought here did not retain their African cultures and languages intact. Instead they formed a new Creole society, a melting pot where ethnic divisions dissolved into a culture more Afro-Caribbean than African. As a consequence São Tomé e Príncipe does not suffer from the ethnic violence that plagues the African continent, and its parties cannot be distinguished by their ethnic background. Rather, the 140,000 inhabitants of São Tomé, and the 5,000 inhabitants of Príncipe, share a language, culture and history. Moreover, after 500 years of Portuguese colonialism—one of the longest periods of European domination in colonial history—the peoples of these islands share profoundly deep family and kinship ties. Locals say that the abbreviation of São Tomé e Príncipe, STP, *means Somos todos primos*, or "We are all cousins."

Multiparty democracy was introduced to the archipelago after the Cold War, and has been marked by continuous political instability, weak state institutions, and widespread corruption. Although it does not suffer the ethnic conflict that plagues mainland Africa, a handful of rival family groups have created real partisan cleavages that define electoral politics on the islands. As Seibert demonstrated in his definitive 615-page *Comrades, Clients and Cousins* (2006), political parties in São Tomé cannot be distinguished by their ideology, statutes, or programs, nor by the ethnic or social backgrounds of their leadership or their following. Instead, "Party politics is primarily based on personality and divergent group interests. The political parties can be distinguished by the history of their leaders and their mutual and personal conflicts, which in turn constitute the *raison d'être* for the parties' existence." (Seibert 2006: 441)

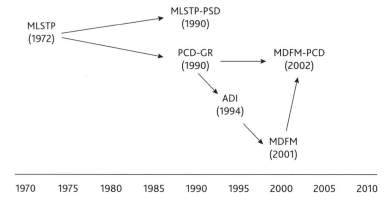

Figure 7.1 Origins of political parties in São Tomé e Príncipe

The electoral law of 1991 declares that the National Assembly has 55 deputies who are democratically elected by proportional representation, with votes turned into seats according to the D'Hondt method. Proportional representation tends to produce multiparty systems, and this is what happened in São Tomé e Príncipe after the introduction of its 1991 electoral law. What had been a one-party socialist state was suddenly transformed into a multiparty liberal democracy. Currently five parties are seated in the National Assembly. Using the Laakso and Taagepera (1979) formula ($E = 1/\Sigma p^2$) the number of "effective parties" (E) now stands at 2.9.

But since these parties are not ideologically based, but rather are based on the personalities and rivalries of their leaders, elections have been organized around vote-buying and promises of patronage, rather than ideological programs or policy platforms.

As shown in Figure 7.1, the origins of the party system in São Tomé e Príncipe can be traced back to the *Movimento de Libertação de São Tomé e Príncipe* (MLSTP) which was founded in 1972 by exiled *forros* (the dominant social class on the islands, translated as "freemen," indicating their historical identity as emancipated slaves). The MLSTP was recognized by the Portuguese as the sole legitimate representative of the people of the islands in 1975. In that year, after the right-wing dictatorship fell in Portugal and the islands won their independence, the MLSTP leader, Manuel Pinto da Costa, returned home and, aged only 37, became Africa's youngest head of state. The MLSTP adopted the one-party socialist system of the Soviet bloc, and was identified with international communism. But really the outward modern appearance of a revolutionary party belied its traditional political culture. "The world was bipolar," explained Pinto da Costa many years later. "NATO had supported Portugal, so we got help from the other side. They called us communists. But it had nothing to do with communism." (Shaxson 2008: 150)

The first split of MLSTP into rival factions was between President Manuel Pinto da Costa (1975–91) and Prime Minister Miguel Trovoada (1975–79). Although the endless competition for power between these two individuals is the defining feature of national politics in the country, they started their careers as comrades. Their parents had been close friends, Trovoada's parents were witnesses to the marriage of Pinta da Costa's parents, and one of Trovoada's uncles was Pinto da Costa's godfather. They had been classmates in the mid-1940s at the primary school in São Tomé. As there was no secondary school on São Tomé, both were sent in 1950 to a seminary in Luanda, Angola. Both later went to the Escola Portugália in Lisbon, after which Pinto da Costa left for France (to avoid military service in the Portuguese army) and went on to study in East Germany where he earned his master's (1968) and doctorate (1971) in economics. Meanwhile Trovoada went to France, then Guinea-Conakry, and eventually settled in Libreville, Gabon, where he worked as a teacher at a grammar school through the remainder of the colonial era. (Seibert 2006: 89–99) In September 1960, while spending school holidays together with some of their expatriate friends, they founded an organization that would later become the MLSTP. It called for the complete independence of the islands, the abolition of all privileges for whites, the establishment of a republican, democratic, secular, and anti-imperialist regime, the prohibition of forced labor, and an eight-hour workday, free medical care, and compulsory primary education. (ibid.: 89)

Because of the political situation on the islands, the Portuguese secret police did not allow this nascent nationalist movement to agitate in the colony. Therefore its members lobbied for independence from exile. Trovoada became party leader, and because he was closer to the islands, set up headquarters in Libreville. Gabon was under the neocolonial system of France, and France was opposed to the radical armed struggles for independence in Portuguese Africa. Trovoada adopted a non-revolutionary, pro-capitalist stance, so his movement was tolerated, and he cultivated life-long ties with the French. (He is sometimes accused of being their puppet.) But it was his lavish lifestyle that particularly caused him problems. After accusations of corruptly using party funds to buy a house and automobile, radical members of the movement removed him from his leadership position in 1965, at which point the party fell under the control of an ideologically radical group known as the *cívicas*. These radicals were the ones who first mobilized the peoples of the islands during the twilight of Portuguese colonial rule, and who elected Pinto da Costa as their leader in 1972, and brought him back to the islands to become the country's first president in 1975.

But Pinto da Costa was not as radical as these *cívicas* activists, so he worked with Trovoada's moderate faction to eliminate them, thereby taking control of the party. Formerly the number one man in MLSTP, Trovoada now became number two, named as prime minister under his old friend—now president—Pinto da Costa. You don't have to be Shakespeare to understand the psychology of rivalry that emerged. The MLSTP established a one-party socialist regime along the model of communist parties in the Soviet bloc, where the president was appointed by a legislature that was itself entirely controlled by a single party. As time went on Pinto da Costa became more autocratic, replacing his older comrades with individuals personally loyal to himself. In 1979 he removed his old friend (now rival) Trovoada from office—and to prevent anyone else from challenging his personal power, he also eliminated the office of prime minister itself.

Kinship was central to the creation of personal rule under Pinto da Costa and the MLSTP. His brother became a minister. His paternal half-brothers became ministers, and directors of the national fuel company and the largest cocoa plantations. His first cousins became party secretaries and ambassadors to Angola. His wife ran the national Red Cross. Her sister became minister and president of the National Assembly, whose brother became minister, as did her first cousin, while her husband became prime minister when the office was finally re-introduced in 1988. (Seibert 2006: 498–9)Although this would appear to be sufficient evidence of nepotistic corruption, and was constantly denounced as such by his rivals, in reality Pinto da Costa was simply using a traditional cultural pattern left over from the colonial era that was even employed by his rivals when they came to power. "The phenomenon of favoritism based on kinship ties and the predominance of one family in the government is locally referred to as *familiarismo*." (ibid.: 497)

The regime was tied to the Soviet bloc, and thanks to the support of those allies a one-party authoritarian regime was maintained where there was no place for open debate or opposition. Although the regime was never as violent as other similar regimes, political dissidents were intimidated and persecuted by a secret police, and many fled into exile (notably Trovoada and his supporters, who maintained close ties with France). Pinta da Costa established a personal dictatorship that relied increasingly on the external support of Angolan and Cuban patrons. Nationalization of the economy failed to generate economic growth, and the cocoa plantations, which provided over 90 percent of export revenues, fell into decline. By the 1980s its poor economic performance convinced the regime to abandon state-led socialist development, and MLSTP finally approached Western donors for financial aid. In 1984, Pinta da Costa signed his first agreement with the IMF and started the country on the path of liberalization. Under donor pressure, São Tomé became the first lusophone country to abandon the one-party state. In 1989, Pinto da Costa opened up space for the creation of opposition parties, and allowed former dissidents to return and participate in the political system.

This was how the *Partido de Convergência Democraticá* (PCD) emerged in the newly created political space in 1990. Originally a broad coalition uniting almost the entire internal opposition to the Pinto da Costa regime, the PCD was dominated by former MLSTP *cívicos* who had returned from abroad during the period of economic liberalization. Many had been members of the old regime before being purged during Pinto da Costa's autocratic phase. But despite their MLSTP origins, they were a younger generation of politicians who were tightly united by their common experience of dissidence and exile. When the first truly democratic multiparty elections were held in 1991, the PCD soundly defeated the former ruling party (renamed MLSTP–PSD to denote its alliance with the Social Democratic Party of Portugal) and took control of the government. In their common hatred for the president, they supported Miguel Trovoada in the presidential elections of 1991, because he personified the opposition to Pinto da Costa. This proved to be a strategic mistake, as President Trovoada (1991–2001) quickly began to resemble his nemesis.

Despite having come to power on accusations of MLSTP corruption, maladministration and incompetence, the new government led by the PCD in 1991 soon adopted the political attitudes and habits of their predecessors. They considered political power as a means of wealth appropriation and distribution rather than the exercise of public service. Kinship was equally important. Daniel Daio (1991–92), the first PCD prime minister, was married to the sister of the brother of the second PCD prime minister, Norberto Costa Alegre (1992–94), whose wife Alda Bandeira was minister of foreign affairs. His younger brother Filinto Costa Alegre was head of the parliamentary group in the National Assembly; Filinto's wife's first cousin Armindo Aguiar was minister of labor. Costa Alegre's first cousin, José Luís Neto Alva, was head of the prime minister's office. (Seibert 2006: 499) Core PCD leaders were kin and peers of each other who grew up in the same neighborhood. Their parents were friends. They went to the same secondary school and began their university studies in Lisbon around the same time in the early 1970s. They returned together as *cívica* activists in the struggle for independence, and after losing their power struggle to Pinto da Costa and Trovoada, had left together for exile in Mozambique. They returned together, and ruled together.

The political honeymoon between this tightly knit PCD parliamentary majority and the president did not last long. The problems began when both the president and the government tried to question the constitutional powers of the other. The first battle ended with Trovoada's dismissal of Prime Minister Daio in 1992. The PCD then named Costa Alegre as prime minister, and he tried to deprive President Trovoada of his important control of foreign aid by removing the cooperation department to the ministry of foreign affairs. Trovoada retaliated by forming a tactical alliance with his old enemies in the MLSTP, who were now the largest opposition party in parliament, scheming with them to pass a vote of no confidence that brought down the PCD Costa Alegre government. When the PCD leader Evaristo Carvahlo accepted

President Trovoada's nomination as prime minister, he was expelled from the party, which had become a political party defined by its personal animosity towards President Trovoada. In 1994, lacking the support of his own party, Carvahlo was replaced by a new MLSTP government, ending the first period of alternation of power in the country's fledgling multiparty democracy, and returning the MLSTP to government. Carvahlo was expelled from the PCD when he accepted the prime ministership.

Table 7.2   Presidents and prime ministers of São Tomé e Príncipe

President	Prime Minister	Party	Government
Manuel Pinto da Costa (1975–91)	Miguel Trovoada	MLSTP	1975–79
	Celestino Rocha da Costa	MLSTP	1988–91
Miguel Trovoada (1991–2001)	Daniel Daio	PCD	1991–92
	Norberto Costa Alegre	PCD	1992–94
	Evaristo Carvahlo	†	1994
	Carlos Graça	MLSTP-PSD	1994–96
	Armino Vaz d'Almeida	MLSTP-PSD	1996
	Raul Bragança	MLSTP-PSD	1996–99
	Guilherme Posser da Costa	MLSTP-PSD	1999–2001
Fradique de Menezes (2001–10)	Evarista Carvahlo	ADI	2001–02
	Gabriel Costa	ADI	2002
	Maria das Neves	MLSTP-PSD	2002–04
	Maria do Carma Silveira	MLSTP-PSD	2004–05
	Tomé Vera Cruz	MDFM	2006–08
	Patrice Trovoada	ADI	2008
	Rafael Branco	MLSTP-PSD	2008–

† Carvahlo was expelled from the PCD when he accepted to be prime minister

It was during this 1994 power struggle that the *Acção Democrática Independente* (ADI) was founded by followers of President Trovoada who splintered off from the PCD and gained substantial support in the 1994 legislative elections, eventually winning several cabinet posts in the 1996 MLSTP government. One of these new ADI deputies was Fradique de Menezes, a rich businessman with many personal connections in Western donor countries. He would later become president, and found his own party.

The ADI was entirely based on the personality of Miguel Trovoada and was unthinkable without him. "It was a political party that depended largely of the personal and political interests of one single person, who used his external ties with France and Gabon to sustain the party's existence. Therefore, during his presidency there was almost no competition for party posts." (Seibert 2006: 295) As Trovoada's second presidential mandate came to an end, and he was no longer in a position to offer them patronage, many ADI deputies could see that their access to patronage and personal privileges would disappear. Then the president imposed his son Patrice Trovoada as party leader, in the usual pattern of kinship–family power. This led to the defection of many ADI leaders, and so the Trovoadas had to find someone loyal who could run under their party banner in the presidential elections of 2001. They settled on Fradique de Menezes, a businessman who had been educated in Portugal and Belgium and who had worked for a series of large international firms, including ITT, Archer Daniels Midland, Goodyear, and Memorex, before returning home in 1986 and becoming the country's top cocoa trader and cement importer. (Shaxson 2008: 149) He had served in the final MLSTP governments as foreign minister and then cooperation minister under Pinto da Costa before being expelled from the ruling party in 1992. It was then that he joined Trovoada's ADI, and was hand-picked by the father and son to run for the presidency, which he won in the first round against a rehabilitated Pinto da Costa.

Initially Menezes was considered a subordinate of his political godfathers the Trovoadas, but as soon as he became the new president, he quickly revealed his own personal ambitions. The *Movimento Democratico Forças da Mudança* (MDFM) was formed when Menezes asked his followers to create a political party to defend his interest in parliament. Carlos de Neves and his faction abandoned the ADI and joined the MDFM soon after Patrice Trovoada announced that he was going to take over the party leadership. They quickly assumed leading positions in the MDFM, which some critics called the "Movement for the Defense of Fradique de Menezes." But in the same pattern of struggle for power between the president and legislature, when Menezes dismissed the government and called for new legislative elections in 2002, the members of his party stopped supporting him, formed an alliance with the opposition PCD, and proposed constitutional amendments to curb presidential power. (Thereafter the coalition has been known as MDFM–PCD.)

The disadvantages of multiparty systems are well known to political scientists. They suffer from chronic instability, because no party tends to hold a majority of seats and they must form coalition governments. Long negotiations between the coalition parties for ministries and cabinet portfolios often reduce their effectiveness. These negotiations by party elites are also out of the reach of ordinary voters. Furthermore, multiparty coalitions are not as accountable as majority party governments, as the responsibility of their individual parties is obfuscated. Classical political scientists such as Marice Duverger (1954) blamed proportional representation and multiparty systems for the lack of ideological moderation that eventually led to the breakdown of democracy in most continental European countries in the 1930s. In São Tomé, where kinship and patronage have replaced ideology, two decades of true multiparty democracy have resulted in widespread rent-seeking, endemic nepotism, and highly corrupt misappropriation of government revenues.

OIL AND CORRUPTION IN MULTIPARTY DEMOCRACY

Is multiparty democracy feasible in African oil states? The experience of São Tomé shows that it is. Not only have there been consecutive multiparty elections, judged fair and free by most outside observers, but the country has also passed the famous "double-turnover" test of democratic consolidation, whereby a party that takes office after a democratic election should relinquish office after losing a comparable election without seeking to resist or overturn the result. (Huntington 1991: 266–7) Despite the two short-lived bloodless military coups by disgruntled soldiers in 1995 and 2003, there is no question that democracy has become the only game in town, and the country has consolidated its democratization process. In fact, in the country's first short-lived bloodless military coup in 1995, the junior officers who briefly arrested President Trovoada called their action a "democratic coup," and showed no ambition for holding power. But can this multiparty democracy fight the corrupting influences of oil? The soldiers who led the 1995 coup blamed Trovoada for their poor conditions (for instance, monthly salaries of $14), but also accused the politicians of enriching themselves without any scruples at the expense of the people. Second Lieutenant Manuel Quintas de Almeida said he and his cohort had been compelled to take control of the country because it had become ungovernable "as a result of a lack of institutional loyalty, seriousness, and transparent management of public funds due to the selfishness and individualism of some Santomeans who have forgotten their country and their people." (Seibert 2006: 258)

When international oil companies first started negotiations for exploration licenses in the offshore waters of São Tomé—containing an estimated 7 billion barrels of oil—nobody in the archipelago had any know-how or experience in the oil sector. Nepotism had produced a competency deficit where those who knew did not rule, and those who ruled did not know. The government granted an unknown Louisiana-based oil waste disposal cleanup company, Environmental Remediation Holding Corporation (ERHC), exclusive rights to the country's offshore reserves. ERHC had entered into negotiations with Prime Mininster Raúl Bragança in 1996. Bragança had no knowledge of the oil business—he was in power because of his kinship ties, not his training or experience. But in 1997 he accepted a $2 million check from ERHC in exchange for rights worth billions, and named his nephew, Carols Bragança Gomez, as the president of the new national oil company, the *Sociedade Nactional de Petróleo de São Tomé e Príncipe* (STPetro). Bragança had come to power through fair and free democratic elections, but that did not make him any more capable of representing his country's interest, nor any less corrupt.

The ERHC contract, when it was revealed through São Tomé's small-town rumor network, became an instant scandal, and was unilaterally rescinded in 1999 by the next government of Guilherme Posser da Costa. The ERHC needed a bold new strategy, and found it in an influential Nigerian businessman, Emeka Offor, who helped to finance Menezes' presidential campaign in 2001. Offor bought a controlling interest of the ERHC for $6 million (Shaxson 155) and only six days after the sale, Nigeria and São Tomé mysteriously settled their offshore territorial dispute over the oil-rich waters, creating a Joint Development Zone (JDZ) in which Nigeria held 60 percent and São Tomé 40 percent

The JDZ is an offshore region located in the territorial waters of Nigeria and São Tomé, based on a 2001 treaty between the two countries. It frames the legal conditions for the joint exploration of the disputed area, and established a Joint Development Authority (JDA) in charge of granting licenses, collecting revenues, and resolving conflicts and disputes involving the zone. On the surface the JDZ looks like a diplomatic solution to the problem of two countries with conflicting claims to the same waters under the Law of the Sea. But beneath the surface it is a scandalous appropriation of a small island's natural resources by a larger regional power. Nigeria has a population one hundred times that of São Tomé, and possesses unchallenged military superiority. The vastly unequal power of these two countries explains why São Tomé—in whose waters the JDZ is located—was forced to accept unfair terms compared to its larger neighbor. President Trovoada had managed to negotiate his country's maritime boundaries with Equatorial Guinea (1998) and Gabon (2001) based on the customary principal of equidistance (where a geometric median line is drawn between two countries as their border). But when it became clear that the international oil industry was going to start drilling in the offshore waters of São Tomé, Nigeria refused to offer the same terms, and instead demanded a "negotiated settlement," which, because of its preponderance of power, resulted in an unfair annexation. The entire JDZ lies on São Tomé's side of the geometric median line.

When President Menezes came to office in 2001 he was advised by the IMF to have an American lawyer look at the new agreement with the ERHC (now owned by Chrome), which had replaced the earlier 1997 agreement. In this new agreement Offor had acquired 15 percent in two offshore blocks of his choice, a 5 percent share in the signature bonuses, a 10 percent share of oil profits, a 1.5 percent royalty from production, two additional blocks of his choice in the country's exclusive economic zone offshore (without having to pay any signature bonuses), and 15 percent in another two blocks. Keep in mind that the ERHC had no other assets except its São Tomé rights, and was not even a real oil company capable of deepwater oil exploration. Moreover, the agreement was binding for 25 years. The World Bank lawyer said that the contract was grossly unfair, so Menezes wrote to ERHC/Chrome saying that the 1997 contracts would be terminated. In response Offor threatened to ask Nigerian President Obasanjo not to ratify the JDZ treaty and to impede oil exploration for many years to come. This was how President Menezes was pressured into signing both agreements: a 60–40 JDZ agreement with Abuja that was clearly unfair, and an agreement with Chrome that foreign experts consider one of the worst in the recent history of Africa's oil history. (Frynas, Wood and Soares de Oliveira 2003: 67) Menezes had been democratically elected in multiparty elections, but this did not enable him to resist foreign pressures by the regional superpower and its Nigerian "businessmen."

The prime minister's nephew, Carlos Bragança Gomez, was named president of the country's national oil company, STPetrol. He was also appointed to the JDA. Besides this ordinary nepotism, he was also involved in a serious conflict of interests, for Gomez was hired by ERHC/Chrome as a "consultant." This meant that he represented the government, the JDA, and a bidding company in the upcoming licensing process for offshore oil concessions. The first licensing round was held in 2003. Twenty companies submitted bids for eight out of nine blocks in the JDZ. With the exception of the American majors Exxon, Chevron, and Andarko, which bid on Block 1, the rest of the bids were made by totally unknown Nigerian "businessmen," in speculative real estate operations whose intention was not to explore or drill for oil, but simply to re-sell their blocks to real oil companies at a profit. For instance, two Nigerian companies, Foby International and ECL International, bid on Blocks 2 and 4, respectively. They had no known oil operations anywhere, and certainly did not possess the offshore rigs required for deepwater drilling in the Gulf of Guinea. Soon after the bidding, it was also announced that the whole process could not be concluded until ERHC/Chrome was allowed to exercise its pre-emptive (bonus-free) options, which it did. The company owned by Emeka Offor exercised four signature bonus-free options in Blocks 6, 3, 4, and 2, and took another two stakes, with signature bonuses, in Blocks 5 and 9. "The four signature bonus-free options would cost STP almost $75 million in lost income," noted Seibert, "or 125% of the country's GDP." (2005: 241)

This situation was intolerable for those not privileged with the patronage and kickbacks, and so on July 16, 2003, while President Menezes was at a conference in Nigeria, disgruntled soldiers launched another coup. This time half the country's 400 soldiers participated. They denounced their poor conditions and the endemic corruption by their country's politicians. Fortunately for Menezes, his collaboration with Nigeria enabled him to get the immediate military support of President Obasanjo, who flew into São Tomé to pressure coup plotters to hand back power. Unable to withstand a threatened military invasion by Nigerian forces, they really had no choice, and negotiated for a general amnesty for themselves, with a promise of several months' unpaid salaries. Some experts have suggested that the 2003 coup was inspired by Nigerians who wanted to send a message to President Menezes that he should not become too nationalist in his oil dealings. Whatever the case may be, after being reinstalled in power, Menezes was forced to accept Patrice Trovoada, his scheming political adversary, who was by then a Nigerian ally, as special oil adviser. "We are bombarded with seminars, conferences, and lectures on transparency and good governance," complained Menezes. "But when the Nigerians threaten, who will call Obasanjo and tell him to lay off? Tony

Blair? George Bush? [...] Will Transparency International or Global Witness ring up? No! They are just criticism programs. Meanwhile my neck is on the line." (Shaxson 2005: 162)

In 2003 it was accidentally revealed that Patrice Trovoada had signed a deal with the Guernsey-based Energem Petroleum, giving the company 70 percent of all profits from the re-sale of crude from other African producers. This deal had been signed without the knowledge of the president or the prime minister, resulting in a political crisis. When the Energem scandal was revealed, Menezes received outside advice from Columbia University's Earth Institute, headed by economist Jeffrey Sachs and financed by billionaire George Soros, who had arrived in 2003 and had offered their services free of charge. A team of Columbia lawyers drafted a new oil revenue management law for São Tomé, with the collaboration of the World Bank, passed by the National Assembly in 2004. It established new management principles such as auditing procedures, transparency regulations, and funds-for-the-future: a typical model of the type of "good governance" then demanded by non-governmental organizations.

Unfortunately, the second licensing round of 2005 did not go much better than the first. No major oil company participated. Exxon declined to exercise two of its options because it was unable to get operational control from Nigerian firms such as Momo Oil & Gas and Equinox Petroleum, owned by Mohamed Asebelua, a Nigerian "businessman" who was also Menezes' special adviser on foreign investments. Another company registered in British Virgin Isles, Equator Exploration, owned by a Canadian national, Wade Cerwakyo, who had business ties with Patrice Trovoada, also bid on blocks. But when Equator received a smaller share than Trovoada (a reported shareholder) had desired, Menezes used the occasion to sack his special oil adviser for conflict of interests. Since Patrice Trovoada was the head of the ADI in the National Assembly, this resulted in another political crisis, with the ADI minister of natural resources resigning from office. Meanwhile ERHC won bids with its consortium partners (real oil companies) in Blocks 2 and 4. An unknown Nigerian–Iranian consortium obtained operatorship of Block 5, and another unknown Nigerian firm became the operator in Block 6. In total, five signature bonuses generated $283 million in the second round of licensing in 2005. Unfortunately São Tomé got only 40 percent of JDZ revenues, received nothing from ERHC's signature bonus-free rights, and also had to pay for expensive new JDA headquarters in Abuja, as well as pay back $8 million that

Nigeria fronted to set up the Nigerian-based JDA and to employ its Nigerian staff. The amount of money received by the government of São Tomé, in the end, was only $57 million.

The offshore waters of São Tomé are rich with oil, but they are also full of sharks. What is a poor, small, weak country supposed to do when powerful neighbors take away its territory and appropriate its oil revenues? Oil wealth continues to attract United States interests and cooperation. In early 2002 the African Oil Policy Initiative Group, a lobbyist for the US oil industry, advocated the establishment of a military base in the islands. Top military brass visited the archipelago, and while denying they were interested in building a naval base, nevertheless began military cooperation with the oil-rich archipelago. Soon the US Trade and Development Agency granted $800,000 to finance a feasibility study for the creation of a deep-sea port. The US military destroyed the obsolete arms supplied by the USSR in the late 1970s. A Millennium Challenge Account was also opened for the country because of the "archipelago's commitment to political reform in favor of democratization and economic liberalization." (Seibert 2004: 244–5) One of the differences between the MLSTP–PSD and the MDFM has been this shift in alliances. While Pinto da Costa's party is allied with Portugal, Angola, and China, Menezes has forged new alliances with the United States, Nigeria, and Taiwan. In July 2005 a US coastguard cutter paid a visit to the islands and conducted joint exercises with the local coastguard. In 2006 the US navy donated a patrol boat to the coastguard, and it has selected São Tomé as a centre for its regional radar program. In 2007 Menezes was flown by helicopter to visit a US guided-missile frigate, where the vice-admiral promised American support in patrolling the country's waters. Meanwhile US oil companies ChevronTexaco, ExxonMobil, and Andarko all acquired major offshore oil concessions during this same period.

Multiparty democracy is something good in itself, but it is not sufficient to protect a small, poor, and weak country from big foreign powers. When I think of São Tomé I am reminded of the famous Melian Dialogue in Thucydides' *Peloponnesian War*. For those not familiar with this story, let me explain. The war between Athens and Sparta had polarized the Greeks into two large alliances, but the island of Melos wanted to remain neutral, and had refused to join either side. So Athens sent 30 warships whose envoys made proposals to the Melians, threatening them with destruction if they did not join the alliance. The Melians argued for neutrality, for justice, and for the ideals of humanity. The Athenians argued for alliance, for war, and for the realism of power. In one of the most famous lines from this dialogue the Athenian delegate said that, "the powerful exact what they can, while the weak yield what they must." (V. *lxxxix*) After a long debate between idealism and realism, the Athenians made a final demand, which the Melians refused. So the Athenians thereupon slew all the adult males whom they had taken, and made slaves of the children and women. Reasoning by analogy, São Tomé is like Melos, and Nigeria is like Athens. São Tomé can make all the idealistic arguments of justice that they like: that the oil is their sole and unique resource capable of bringing them out of five centuries of poverty, that it is located on their side of the line of equidistance, that they are a peaceful neighbor, that they are a multiparty democracy. In the end Nigeria is bigger and stronger, and the powerful will take what they can, while the weak will yield what they must.

8
Armed Struggle for Independence

OIL AND WAR

The idea of fighting over African resources is not new. The colonization of Africa was driven in part by a violent quest for valuable commodities. Today conflicts over natural resources are increasingly frequent, driven by the relentless expansion in global demand, the emergence of significant resource shortages, and the proliferation of ownership contests. "Disputes over access to critical or extremely valuable resources may lead to armed conflict," a special kind of violence that Michael Klare calls "*resource wars.*" (Klare 2001: 25) Of all the natural resources, none is more likely to provoke conflict between states in the twenty-first century than oil. "Petroleum stands out from other materials—water, minerals, timber, and so on—because of its pivotal role in the global economy and its capacity to ignite large-scale combat." (ibid.: 27)

Klare's principal concern is the potential for inter-state wars, especially the likelihood that the world's major powers will go to war over natural resources. But most of his actual cases are domestic intra-state wars in the developing world where natural resources such as oil, timber, or diamonds are concentrated in an area occupied by groups seeking to break away from the existing state. "Such contests are regularly described in the international press as ethnic and sectarian conflict," he objects, "but it is the desire to reap the financial benefits of resource exploitation that most often sustains the fighting." (ibid.: 190–1) Denied access to political power, in an economy controlled by a ruling faction or family, such groups often see no option but to engage in armed rebellion. "Once a rebellion has erupted," he explains, "the fight often evolves into a resource conflict." (ibid.: 193)

This raises an important question. If the conflict comes *before* the struggle for resources, then are natural resources really the cause of the war, or merely another aggravating factor? Klare was influenced by the writings of Paul Collier (1998, 2000), an economist who links the outbreak of internal resource wars to greed and economic opportunism, rather than to larger structural inequalities or deep-rooted grievances. In his recent bestseller Collier reaffirms this idea that low income and slow growth make a country prone to civil war: "Why? Low income means poverty, and low growth means hopelessness. Young men, who are the recruits for rebel armies, come pretty cheap. Life is cheap and joining a rebel movement gives these young men a small chance of riches." (Collier 2007: 20) Collier's famous formula is that, "Grievance has evolved, over the course of the decade, into greed." (ibid.: 30–1) Klare similarly reasons that, "With so much at stake, and so few other sources of wealth available in these countries, it is not surprising that ruthless and enterprising factions are prepared to provoke civil war or otherwise employ violence in the pursuit of valuable resources." The government, for its part, "is just as likely to fight for these resources, both to pay its bills and to ensure the continued loyalty of prominent cliques and families." (ibid.: 193) African resource wars, in other words, are essentially about *greed*.

Many critiques have been leveled against "greed theorists" such as Collier and Klare who ignore the genuine grievances of rebels, and attribute armed struggles as a whole to greed. Not only can individuals combine the eternal human motives of greed and grievance, but some individuals will be more motivated by one than the other. It is furthermore obvious that some rebellions are motivated by a struggle against oppression, marginalization, and exploitation, while others are motivated by what could be called an entrepreneurial spirit. Instead of holistically attributing a single motivation to an entire struggle, therefore, it is better to speak of the numerous factors that move disputes towards or away from conflict. This avoids, among other things, the fallacy of the single-factor explanation. But it also avoids blaming civil wars solely on the motivations of domestic actors (ignoring the interests and motivations of foreign actors).

Recent research published by highly regarded scholars and leading international financial institutions such as the World Bank have shown that developing economies with high rates of dependence on extraction and export of natural resources do have a correspondingly high propensity to violent conflict. (Ross 2003) High dependency on oil *is* correlated with war. But what such large-n, quantitative, correlations between oil and war fail to address are the many qualitative differences between resource conflicts. Some are armed struggles about ownership and control over resources which could be called "resource wars." But others are struggles over the distribution of revenues derived from natural resources. These are not resource wars, but "revenue conflicts." Some are about the inability of weak state institutions to cope with looting, misappropriation, and exclusion of significant sectors of society, leading to violent protests. These are not wars, but domestic "police matters" of maintaining public order. Others are about states using their resource revenues to build up repressive security machinery and embarking on violent terror against their own people. These are not wars, but one-sided "violent tyrannies." Some are illegal uses of resource revenues by disgruntled factions of the governing elite to sponsor anti-government insurgencies or secession movements. These are not wars, but "factional politics" using violence as leverage. Others are organized predation and extortion of big businesses in the resource extraction sector by aggrieved groups. These are not wars, but "organized crime." Some are military interventions by foreign stakeholders to protect their investments. These are not (called) wars, but "peacekeeping operations." (Omeje 2008: 14–15)

This chapter concerns itself with one kind of resource conflict, internal armed struggle for regional independence, in which natural resources are not the cause, per se, but only an aggravating factor. (The next chapter will deal with conflicts where oil is the cause.) It addresses the following questions: Can Africans living in an oil-rich region emancipate themselves from "violent tyranny" by means of armed resistance when that regime is supported financially, diplomatically, and militarily by foreign powers? When foreign powers crave their oil, when international governance initiatives prove insufficient, when their states are unwilling or incapable of changing themselves, when opposition parties lack democratic elections, when the press is not free, then can armed struggles succeed in fighting their "paradox of plenty" from below?

Table 8.1 shows that most armed struggles for independence of oil-rich regions of Africa have failed to achieve their goals of self-determination. After four decades of low-intensity conflict by the FLEC in the Cabinda Enclave (formerly known as "Portuguese Congo") the native Kongo people failed to emancipate themselves from the military regime in Angola. Similarly, the UPC guerrillas who fought for the English-speaking peoples of the Western Region (formerly "British Cameroons") were defeated militarily by the French-backed central government. The indigenous Bubi people of

Table 8.1   Selected armed struggles for self-determination in oil-rich regions of African states

Angola	Cabinda Enclave Liberation Front (FLEC)	1963–2006	Failed to achieve independence of Cabinda Enclave
Cameroon	Union of the Populations of Cameroon (UPC)	1948–71	Failed to achieve autonomy of the Western Province
Equatorial Guinea	Movement for the Self-Determination of Bioko Island (MAIB)	1994–present	Failed to achieve independence of Bioko Island
Nigeria	Republic of Biafra	1967–70	Failed to achieve independence of Eastern Region
Sudan	Sudan People's Liberation Movement & Army (SPLM/A)	1983–2005	Achieved legal autonomy of Southern Sudan

Bioko island (formerly "Fernando Poo") struggled for independence from Equatorial Guinea and were massacred by the Fang regime, and today over two-thirds of them live in exile, where they run their underground movement. In the Nigerian civil war the federal government crushed the rebellious Igbo people, who declared their independence as the Republic of Biafra (formerly "Oil River States"). While these regions are oil-rich, it would be unfair to say that their armed struggles were motivated simply by greed for oil. These were genuine liberation struggles.

They were different from other African struggles, whose goal was not regional secession, but overthrowing a regime in power. The numerous civil wars in Chad were never about achieving regional independence from Ndjamena, but about overthrowing its corrupt rulers. The same is true for the rebel movements in Angola and Congo. Neither Angola's UNITA nor Congo's "Ninjas" were about regional secession—they were rather about national unity under a new regime. What is interesting is that the only successful armed struggle for regional self-determination in oil-producing Africa— the SPLM/A of Southern Sudan—came after it changed its strategy from *regional* secession to *national* liberation. Instead of fighting exclusively for the liberation of Southern Sudan, the rebels changed their goal to the liberation of all the people of Sudan. There is a lesson in their victory for other similarly situated armed struggles in oil-rich regions of Africa and the rest of the world.

REGIONAL IDENTITY AND VIOLENCE

What is a region? It always denotes a geographical *space*. But beyond that there are many meanings attached to it. It often has a cultural element. It may contain a distinct society and a range of social institutions. In other words it may relate, more broadly speaking, to an *identity*. "So while the concept of region must always be associated with territorial space, it must be understood as a social, economic, and political construction" which is to say, "the historical work of human actors and actions." (Bickerton and Gagnon 2009: 368) Distinct regional cultures can sustain a sense of regional community and provide a basis for values and policy preferences that differ from other regions or the larger national community. These regional traits are available to be mobilized politically. A region's history, mythology, and cultural symbols may become an ideological resource for political actors. Regional minorities with claims to historical nation status have long contradicted conventional assumptions about cultural assimilation of early modernization and development theorists (Deutsch 1966; Shils 1975) who portrayed regions and regionalism as remnants of pre-industrial, pre-modern societies, fated to be eclipsed by the nation state. Modernization, they thought, would lead to cultural homogenization and an inevitable decline of regionalism.

Modernization theorists were challenged by cultural theorists (Almond and Verba 1963; Hartz 1964) who argued for the persistence of identities. In postcolonial states in Africa, where ethnicity had been part of the colonial project of divide and rule, and where modernization and development failed to occur, there were few signs that ethno-regional identities would simply fade away. On the contrary, their relevance increased over time. (Rothchild 1985; Young 1994) But it is equally true that cultural differences can diminish over time while regional identities persist. In Europe, for example, many regions have lost their cultural distinctiveness yet continue to pursue regional politics in the EU budget. In North America, regional identities became more politicized precisely at a time when inter-group cultural differences were much less significant than they had been historically. (Gibbins 1980) In Africa, Southern Sudan contains many different ethnicities, languages, and religions, so southern regional identity is not strictly speaking cultural. We should treat regional identity and cultural distinctiveness as independent phenomena. Some regional identities cannot be reduced to culture.

An alternative Marxist approach saw regional grievances as stemming from unequal relations between the core and the periphery. Capitalism produces regional disparities and uneven economic development, which structure the rise of minority nationalist movements and set in motion socio-political fragmentation. (Nairn 1977) This is relevant to oil-rich regions in Africa, particularly economic grievances over oil revenues. But like greed theory, it reduces complex historical struggles to economics. In a way it is a kind of anti-greed theory. It doesn't explain, for example, the persistence of regional identities in affluent regions, nor in places where capitalism has not emerged. Economics alone cannot explain everything. Regional identity is a complex historical, social, political *and* economic phenomenon.

Still, regional conflicts based on grievances against central governments in oil producing states *are* a symptom of the "paradox of plenty." (Karl 1997) Onshore regions located above major oil reserves are often the poorest and least developed. This geographical paradox generates economic grievances that combine with other historical, cultural, and political grievances to give conflicts an appearance of "resource wars." But invariably, when we closely examined these struggles, we find that the causes are deeper, that oil is really just fueling the flames of war. Therefore it is better to think of these armed struggles less as resource wars and more as conflicts over *identity*. Not only does this directly address the problem of nation-building, but it also accepts their self-definitions as movements. Now here is the problem. When such armed struggles fight to defend their regional identity, or control the oil or oil revenues of their region, experience has shown that they are unlikely to win. There are many explanations: International law privileges nation-states over rebel armies; oil revenues provide central states with more money than rebels to buy arms and ammunition; foreign oil interests work to protect their investments with foreign security forces; and the international community usually prefers peace to armed struggle.

Even so, the real question is why Southern Sudan succeeded whereas others have failed. It may be that by pursuing purely regional goals of self-determination, those other armed struggles have prevented their own success. Changing strategy could change their outcome. This is where the writing of Nobel laureate Amartya Sen offers a possible solution. In his *Identity and Violence* (2007) Sen argues that a strong and exclusive sense of belonging to one group carries with it perceptions of distance and divergence from other groups. Within-group solidarity helps to feed between-group discord: "Violence is fomented by the imposition of singular and belligerent identities on gullible people, championed by proficient artisans of terror." (Sen 2007: 2) The marshalling of an aggressive Arab Islamic identity in northern Sudan along with the exploitation of racial divisions, resulting in raping and killing, is one of his examples. But sectarian violence across the world, "turns multi-dimensional human beings into one-dimensional creatures:"

> The *illusion of singular identity*, which serves the violent purpose of those orchestrating such confrontations, is skillfully cultivated and fomented by the commanders of persecution and carnage. It is not remarkable that generating the illusion of unique identity, exploitable for the purpose of confrontation, would appeal to those who are in the business of fomenting violence. There is a big question about why the cultivation of singularity is so successful, given the extraordinary naiveté of that thesis in a world of obviously plural affiliations. The martial art of fostering violence draws on some basic instincts and uses them to crowd out the freedom to think. [...] But it also draws on a kind of logic—*fragmentary logic*, (1) to ignore the relevancy of all other affiliations and associations, and (2) to redefine the demands of the 'sole' identity in a particularly belligerent form. (ibid.: 175–6)

To understand the problem that Sen is working on, it is important to note that he is not concerned with the armed struggles for self-determination, but with violence. Sen wants to achieve peace in the world. His own childhood experiences of inter-communal violence between Hindus and Muslim in India led him to seek answers to the problem of the relationship between identity and violence. According to Sen the solution to this kind of conflict is to draw on the understanding that the force of a bellicose identity can be challenged by *the power of competing identities*: "This leads to other ways of classifying people, which can restrain the exploitation of a specifically aggressive use of one particular categorization." (ibid.: 4) For example, he accepts that "culture matters," and can have a major influence on behavior and thinking, other things such as class, race, gender, profession, politics also matter, "and can matter powerfully." (ibid.: 112) Culture is not homogenous, nor does it stand still, but most of all, "culture cannot be seen as independent of other social forces." (ibid.: 113) The same is true for religion. "A person's religion need not be his or her all-encompassing and exclusive identity. In particular, Islam, as a religion, does not obliterate responsible choices for Muslims in many spheres of life." (ibid.: 14) Different persons who are Muslims can vary greatly in other respects, such as political and social values, economic and literary pursuits, their professional and philosophical involvements, attitudes to the West, and so on. "To focus just on the simple religious classification is to miss the numerous and varying concerns that people who happen to be Muslim by religion tend to have." (ibid.: 61)

The illusion of cultural or religious destiny as an independent and stationary force with an immutable presence and irresistible impact is not only misleading, "it can also be significantly debilitating, since it can generate a sense of fatalism and resignation among people who are unfavorably placed." (ibid.: 112) The same is true for what Sen called the "colonized mind," an identity obsessed with the extraneous relation with colonial powers, "hardly a good basis for self-understanding." (ibid.: 89) Sen argues that the devastating effects of humiliation on human lives in Africa left a legacy of fragmented identity. "To lead a life in which resentment against an imposed inferiority from past history comes to dominate one's priorities today cannot but be unfair to oneself. It can also vastly deflect attention from other objectives that those emerging from past colonies have reason to value and pursue in the contemporary world." (ibid.) Decolonization of the mind requires recognition of both endogenous African identities and exogenous Western identities for liberation struggles to counter that nefarious, oft-repeated argument that democracy is a "Western" idea alien to "Africans."

When applied to the regional struggles in Africa, Sen recognizes that simple neglect can be reason enough for resentment, and "a sense of encroachment, degradation, and humiliation can be even easier to mobilize for rebellion and revolt." (ibid.: 144) He knows that "poverty and economic inequality can help to create rich recruiting ground for the foot soldiers of terrorist camps" (ibid.: 145) and that "the neglect of the plight of Africa today can have a similarly long-run effect on world peace in the future." (ibid.: 144) But rather than embracing a singular identity based on one's region, he advocates that people expand their horizons, and share a larger identity with others. In so doing they will not only achieve their goal of liberation, but also liberate their minds from the colonial legacy of divide and rule.

This may sound idealistic, and it is. But as the case study of Southern Sudan will show, such change in consciousness can in fact be a realistic strategy for regional liberation. Extrapolating from Sen's problem of identity and violence to the problem of armed liberation struggles requires the recognition that his pacifist goals are not incompatible with their own. These armed movements do not seek a state of perpetual war. If they mobilize people into armed resistance, it is not, as the greed theorists claim, because they seek personal fortunes, but because they desire certain political objectives. When those objectives are met, they would like the war to stop. The problem the Southern Sudanese are facing now, however, is that, despite the agreement to hold a referendum on the question of the independence of Southern Sudan (2011), it is by no means clear that a vote for independence will result in an end to the long bloody civil war. On the contrary, secession may lead to a return to hostilities. Recognizing that the Northern regime may not be willing to accept Southern secession, in part because of the location of the country's oil reserves, and in part because of the dangerous precedent it would establish for other regions, it has become clear that finding another solution may be preferable. National unity within a larger "New Sudan" may be better for the Southerners than national independence as "Southern Sudan."

In order to achieve unity, however, it will be necessary to change old mentalities that have been shaped over centuries, and have given birth to distinct violent regional identities. Like other attempts to fight the oil curse, what is needed is a *transformation of consciousness*. This transformation is not a rejection of regional identity, nor any of the other identities of ethnicity, language, and religion that have mobilized the people in their liberation struggle, but an inclusion of all their identities, and a realization of the real diversity that underlies the region, the country, and the continent as a whole. It is an ideal encapsulated by the term "multiculturalism," which has its antecedents in the political thinking of Mohandas Gandhi. During his struggle for Indian independence, Gandhi had insisted that while he himself was a Hindu, the political movement he led was staunchly universal, with supporters from all of the different groups in India. When the British sought religious partition of India and Pakistan, Gandhi made a plea for the colonial rulers to see the *plurality* of India's diverse identities. There were the differences of class, of gender, of language, and of regional traditions that would not be resolved by a simple partition along religious lines.

In the end the British partitioned India and Pakistan, but did not resolve religious conflict. One disastrous consequence of defining people by their religious community and giving that single

Map 8.1 Sudan

religious identity predetermined priority over all others was the Indo-Pakistani wars. Meanwhile multicultural India, with a Muslim population nearly the same size as Pakistan, has largely been able to avoid violent indigenous Islamic terrorism. Established democracy, federalism, and multiculturalism are widely accepted in India, and being Indian is more than a religious identity. Despite the efforts of Hindu nationalists, "Ghandi would have taken some comfort in the fact that India, with more than 80 percent Hindu population, is led today by a Sikh prime minister (Manmohan Singh) and headed by a Muslim resident (Abdul Kalam), with its ruling party being presided over by a woman from a Christian background (Sonia Gandhi)," (ibid.: 167) an ideal, perhaps, for a New Sudan.

SOUTHERN IDENTITY AND VIOLENCE IN SUDAN

Southern Sudanese, in early 2011, voted overwhelmingly (99 percent) for secession from Sudan. Their Southern identity began in the nineteenth century, when slave raids began to extend into the South. Until then the peoples of this region had no idea of themselves as being Southern.

It all started with the Ottoman Empire pushed into the southern part of Sudan (1839–81) in search of fresh recruits for its armies. At first it had captured slaves for its armies in the North, but gradually it pushed further into the South. This remote region was largely unknown to the Turko-Egyptian invaders whose aggressiveness met with a violent resistance. The significant aspect of these invaders, including the Arabs, is that while they persistently raided the South for slaves, they never penetrated deeply, nor did they settle.

The Mahdists revolt (1881–98) was initially popular in the South as an anti-Turkish alliance against a common enemy, but when Southerners did not embrace Islam, the Madhists carried out their holy war in close cooperation with the slavers, and, failing to convert, left a legacy of distrust and fear. "The Mahdi's principal supporters were the Missirya and Rezeigal tribes to whom the Madhist revolution promised restoration of slavery." (Johnson 2003: 105) These are the same tribes who recently engaged in what the regime euphemistically calls the "abduction" of Southern women and children, a revival of slavery in its classical form. "From the point of view of Southern peoples there was little to distinguish between the two groups of plunderers and exploiters" (Alier 1990: 12) and so all of the invaders, in the eyes of the Southerners, became seen as Northerners.

Southern identity began as a tradition of resistance to the imposition of alien ideas and customs upon them. The Dinka disdain for the Arabs today is very closely related to the history of slavery. "They lost hundreds of thousands of cattle," reported Major General Titherington in 1927: "Men, women, and children were slaughtered, carried off into slavery, or died of famine. But the survivors kept alive in the deepest swamps bravely attacked the raiders when they could, and nursed that loathing and contempt for the stranger and all his ways." (cf. Deng 2009: 66) According to Dinka Chief Makuei Bilkuei, "They would come with camels and donkeys and mules and guns saying *La Illah, ila Alla, Muhammad Rasul Allah!* That was the way they chanted while they slaughtered and slaughtered and slaughtered." (ibid.: 67) As a result of the Mahdist revolution, the population of Sudan fell from around some 7 million in 1881 to somewhere between 2 and 3 million by 1898. (Daly 1986: 18)

The British colonial government (1899–1956) decided not to abolish slavery hastily, but instead to discourage it gradually. "The

British drew a distinction between the slave trade and slavery, abolished the former and tolerated the latter." (Collins 1983: 374) As late as 1925 Lord Kitchner, the first Governor-General of Anglo-Egyptian Sudan, declared in a memorandum: "Slavery is not recognized in the Sudan, but as long as service is willingly rendered by servants to masters it is unnecessary to interfere in the conditions existing between them." (Idris 2001: 50)

The British eventually stopped tolerating slavery, but only because of the threat to their power that was being posed by rising Arabization and Islamization. Perhaps the most important development in Sudan was the rise of an Arabic-Islamic ideology that posed a form of Arab resistance to British colonial rule. There were even efforts to unite the Sudan with Egypt into a grand Arab republic. It was in reaction to this threat that new Governor-General Sir Harold MacMichael provided a statement of the British position, known as the "Southern Policy," which reinforced a separate Southern identity for the rest of British rule: "The policy of the Government in the Southern Sudan is to build up a series of self-contained racial or tribal units with structure and organization based on indigenous customs, traditional usages and beliefs." (ibid.: 113) Britain's "Southern Policy" also added a new linguistic element to the nascent Southern identity: "Every effort should be made to make English the means of communication among the men themselves to the complete exclusion of Arabic." (Al-Rahim 1969: 245–9) Northerners spoke Arabic; Southerners English.

Behind the Islamic ideology of Sudan lies a racist system, which divides the population into "Arabs" and "Africans" and even divides Muslims into "Arabs" and "non-Arabs," in that order of value. Like all racist systems, the Arab and African labels reflect perceptions rather than realities, since even those who claim to be Arabs are really African–Arab hybrids. Since the Arabs who came to the Sudan were men who married into African Sudanese families, there is an exaggerated pride in Arabism that stems from a deeper inferiority complex associated with the African connection. "Unlike the elites of the Arab world, who do not need to state the obvious, Northerners need to compliment their lack of features in words." (Deng 2009: 62) Their love for being Arab is mirrored by their disdain for Blackness, and since Blackness is visibly evident in their own physique, this implies a deep-rooted and unconscious self-hatred. Humiliated by the Anglo-Egyptian forces, the Sudanese needed psychological reassurance, which they could not find in their

African identity. The color they use to describe themselves is neither white nor black, but *akhdar*, or "green."

After the rise of a Northern nationalist movement for independence of the Sudan, supported by Egypt, the British were forced to hastily reverse their "Southern Policy" in favor of unity. In 1951 a Constitutional Amendment Commission was established to draft a new constitution that would usher the country to independence. Only one member of this commission was a Southerner. Similarly, the British self-government statute of 1953, which called for the "Sudanizing" of the civil service, in fact "Northernized" it. Out of 800 posts, only eight junior posts were given to Southerners. (Deng: 71) As independence approached, Northern domination and a possible return to the dark age of slavery triggered a Southern revolt. The Southern rebellion began in August 1955, four months before independence, as a result of fears that independence would not only result in domination, but could also result in a return to the Arab enslavement of the Africans. The *Anya-nya* first erupted in the Southern town of Torit, triggering a secessionist war that would last for 17 years, kill more than one million people, and force another one million into refuge in neighboring countries. The rebellion was led by the Southern Sudan Liberation Movement (SSLM), whose goal was the sovereign independence of the South.

The Southern struggle against Northern Arab-Islamic domination was viewed by the central government in Khartoum as a matter of law and order. Official policy sought to crush the rebellion and punish its perpetrators, coupled with a policy of forced Arabization and Islamization aimed at eliminating the non-Arab, non-Muslim identity of the South (perceived as a legacy of the British divide-and-rule policy). "Northerners generally assumed that their identity was the national model, and what prevailed in the South was a distortion that the colonists had imposed to keep the country divided." (Deng 2009: 73) But it was Southern resistance to domination and attempted assimilation that proved to be the key factor in the Khartoum's chronic instability, and its rise of military dictators. Military coups took place in 1958, 1969, 1985, and 1989. When Prime Minister Abdalla Khalil handed over power to General Ibrahim Abboud in 1958, the idea was that the military would be better equipped to deal with the war in the South than civilians. But the civilian rulers were just as warlike.

Sadiq al-Mahdi, the great-grandson of the original Mahdi, became prime minister in 1966. His government declared that it would authorize the army and other security forces in the South to do whatever they saw fit for the maintenance of law and order. "This meant in practice that if the Southern guerrilla army attacked a town, all the Southerners within it were suspects and could be killed for not reporting the presence of rebels." (Deng 2009: 75) If the army went outside the town on patrol and was ambushed, all the villagers in the surrounding areas were condemned to death and their villages burned down. Torture centers were established. Massacres were committed. But resistance only grew stronger. An outrage developed among the Southerners against the Arabs that made them believe that Arabs were created by God as human beings, but with different moral attributes: i.e. vile and depraved. It was victimization of women and children that most outraged the Dinka. Bulabek Malith recalled the depravity: "Arabs are bad. Before they kill you, they cut your muscles to make you an invalid who cannot walk. They ask you to grind grain kneeling down naked, and then they put a thorn on the tip of a stick and give it to a small child to prick your testicles as you grind the grain." (ibid.: 118)

The worst atrocities were committed by the government of Mohammed Ahmed Mahjoub, the civilian prime minister in 1965, and then again in 1967–69. His failure to end the war was a major factor that led to the 1969 coup led by Jaafer Mohammed Nimeiri, in alliance with the Communist Party (the only Northern party sympathetic to the needs of the South). Nimeiri was appalled at the atrocities and inhumanity of the war, and after new presidential elections, brought moderates into his government, "for the most part intellectuals who counseled him to end the war in the South and bring Southerners into the alliance in the center." (ibid.: 76) So it came to pass, as a result of the failure of violence to establish law and order, that an intensive peace process resulted in negotiations in the Ethiopian capital, Addis Ababa, hosted by Emperor Haile Selassie. These negotiations resulted in the Addis Ababa Agreement of 27 February 1972, which brought an end to the first phase of the civil war, and granted autonomy—but not independence—to the South. Joseph Lagu, the SSLM rebel leader, declared that he had never been a separatist: "All he wanted was recognition as a citizen with all the rights of citizenship." (ibid)

Unfortunately the peace process fell apart. Libya invaded Sudan in 1976 to help Arab extremists try to topple the Nimeiri regime. Although Nimeiri managed to survive this Libyan-backed coup attempt (with help from the loyal Southern soldiers of his palace guard), his fragile centrist party coalition couldn't resist rising Arab-Islamic fundamentalist ideology. The North was more populous than the South. Electoral arithmetic was strongly in their favor. Nimeiri was a Northerner. In a tragic reversal of policy he suddenly embraced Islamization, and invited the Umma Party and Muslim Brotherhood to join his government. Nimeiri now found himself standing between two opposing regional identities. Mindful that the South might be an obstacle to Northern Islamization, he manipulated internal ethnic differences in order to divide the region, alleging that the Dinka leaders were dominating other ethnicities. His government passed legislation imposing *shari'a* throughout Sudan, and called for the division of the South into several regions. By 1983 a re-born *Anya-nya* movement started mobilizing soldiers for a new armed struggle to resist these Northern policies of assimilation, domination and division. The second phase of the civil war had begun.

JOHN GARANG AND THE "NEW SUDAN"

The second phase of the civil war was more than a continuation of hostilities. It was a different kind of liberation struggle. For if the Southern Sudanese Liberation Movement (SSLM) had been fighting a regional war exclusively for independence of Southern Sudan, the new Sudan People's Liberation Movement (SPLM) and its Army (SPLA) instead chose to fight for the liberation of all the Sudanese peoples from the regime in Khartoum. This change of strategy was accomplished by a heroic Dinka leader, John Garang de Mabior (1945–2005), whose really big idea was that the South should stop thinking of itself as a victim trying to flee from the hands of a violent state and start believing in its own abilities to change the country. On 22 March 1985 he outlined his vision of a "New Sudan:" (1) establishment of democracy, social justice, and human rights, (2) secular nationalism, (3) regional autonomy and/or federalism, (4) radical restructuring of power, (5) balanced regional development, and (6) eliminating institutional racism (see Box 8.1).

None of these goals was based on a singular Southern identity. Rather than conducting another ethno-regional struggle seeking only to preserve historical traditions from the past, Garang based his struggle on ideological objectives that offered a vision of a better future. He defined the aims of the struggle in terms of democracy and human rights instead of rejecting them as being "Western" and not appropriate for "Africa." He defined the aims of the struggle as redressing regional inequalities in the East, the West, and the far North, which ended the false amalgam of all non-Southerners

Box 8.1 Ideological objectives of the Sudan People's Liberation Movement

"We are committed to the establishment of a new and democratic Sudan in which equality, freedom, economic and social justice and respect for human rights are not mere slogans but concrete realities we should promote, cherish, and protect."

"We are committed to solving national and religious questions to the satisfaction of all the Sudanese people and within a democratic and secular context and in accordance with the objective realities of our country."

"We stand for genuine autonomous or federal governments for the various regions of the Sudan, a form of regionalism that will enable the masses, not the regional elites, to exercise real power for economic and social development and the promotion and development of their cultures."

"We are committed to a radical restructuring of the power of the central government in a manner that will end, once and for all, the monopoly of power by any group of self-seeking individuals, whatever their background, whether they come in the uniform of political parties, family dynasties, religious sects, or army officers."

"We firmly stand for putting to an end the circumstances and the policies that have led to the present uneven development of the Sudan, a state of affairs in which vast regions of the East, South, West, and the far North find themselves as undeveloped peripheries to the relatively developed central regions of our country."

"We are committed to fight racism which various minority regimes have found useful to institutionalize, and that has often been reflected in various forms and colours, such as the apartheid-like Kacha, a policy under which many poor and unemployed have been forcibly driven en masse to their regions of origin, mainly Western and Southern parts of the country."

Mansour Khalid, ed., *John Garang Speaks* (1987)

as "Northern." He defined the enemy not as Northerners, but as particular "family dynasties" and "political parties" who had monopolized power to the detriments of all Sudanese people (even those in the Center). By redefining the goals of the struggle for liberation, his vision of a "New Sudan" allowed the SPLM/A to build multiregional alliances against a common enemy: i.e. a singular Arabic-Islamic nationalism that had divided the Sudan and caused three decades of civil war.

The armed struggle and the political mobilization of the masses was seen as essential to the creation of a New Sudan, with the SPLM/A serving as the instrument of this transformation. An SPLA spokesman at Bergen, Norway, in February 1989 explained the necessity of both the political and military struggle: "The birth in 1983 of the SPLM/A as a politico-military organization furnishes the Sudanese revolutionary struggle with the armed component required to confront the armed custodian of the minority clique rule." (Deng 2009: 43) The idea, explained this spokesman, was to create a "coalescence of all democratic forces into a single revolutionary tidal wave," which could only be achieved if "the mass political movement and the armed struggle converge." (ibid)

Beginning with President Nimeiri, but intensifying with Sadiq al-Mahdi (1985–89) and worsening under Omar al-Bashir and the National Islamic Front (NIF), who came to power by coup in 1989, then adopted the name National Congress Party (NCP), Khartoum recruited, armed, and deployed tribal militias. These tribal militias were supposed to fight the SPLA, but instead terrorized the civilian populations, particularly the Dinka, from which Garang and the majority of the SPLA armed forces came. He recruited Baggara Arabs of the Rizeigat and Missirya tribes of Darfur and Kordofan, reminiscent of the pattern used by the Mahdist revolution in the nineteenth century, who abducted women and children and turned them into slaves, repeating all of the horrible atrocities that had humiliated the Blacks, and in so doing, hardened "African" resistance. Al Bashir's militias also alienated other regional groups, transforming what could have been a regional war between the North and the South into a total war between the Center and Periphery. Ethnic-based armed struggles arose everywhere against the NCP regime of Omar al Bashir: in the Nuba Mountains and the Southern Blue Nile, Beja uprisings in the Eastern region, Nubian opposition in the far North, not to mention the Zaghawa, Masalit, and Fur rebellions of Darfur.

The Nuba Mountains have been called "the north of the south, and the south of the north." Such frontiers always highlight problems of strict territorial/cultural compartmentalization. Nuba is a geographical or regional term, not a monolithic culture. The Nuba people are located at the North–South border. Being Black Africans, they were kept at the margins of Arab-Islamic domination, but were also excluded from the African-Christian-Animist South. Before the war they were not unified, but were divided between those with a linguistic affinity with the originally non-Arabic speaking Northern Sudanese, and those African Nuba who had no common origin with Northern Sudanese besides Islam and Arabic language. (Salih 2009: 277) Although they were clearly non-Arab, the Nuba had been taught to believe that they were part of the Arab stock, and should identify with the Arabs, and take pride in the Arab-Islamic racial, cultural and religious heritage. During the first phase of the civil war, therefore, the Nuba had been distant sympathizers of the Northern regime, and the Nuba Mountains were not affected by the hostilities.

But in the second phase of the war, when al-Bashir recruited Baggara Arabs into militias, the Nuba endured more oppression than any other group in Sudan. In the first two years of the war, the number of internally displaced Nuba in the capital of South Kordofan state alone exceeded 40,000. A Nuba politician, Yusuf Kuwa Mekki, became SPLA military commander in South Korodfan, and led his people into a war of resistance. The government attacked Kamda, Taroji, Tulushi, Tima (1989) Koaleb, Tira, Shat, Miri Barah, Lima, Otoro, Moro, and Heiban (1990–1), calling for an Islamic *jihad* in the Nuba Mountains, and by 1992 was using heavy artillery. Government forces systematically eliminated all independent Nuba leaders in the towns: chiefs, merchants, teachers, health workers. "In fact," wrote one observer, "anyone with an education is liable to be arrested and tortured, executed or 'disappeared.' The aim is to decapitate communities and leave them without the means to defend their interests." (De Waal 2001: 132) Then the government created death squads for the villages. The fact that the people of the Nuba Mountains and Southern Blue Nile are Muslims shows that this conflict was clearly racial, and not, as the regime pretended, religious.

But internal divisions in the ruling NCP allowed the Southerners to gain the upper hand, and by 1999–2000 the SPLA forces had regained much of its lost territory. Khartoum found itself fighting a war against all of the peripheral regions at once, in a million-square-mile territory that it barely controlled. One of those regions—Darfur—became a symbol of the depravity of the second phase of the civil war. The tragic events that brought Darfur to the forefront of international attention, the barbarities committed by the government-backed *Janjaweed* militia, culminated in a most appalling humanitarian disaster, resulting in the displacement of over two million people, in addition to the 200,000–300,000 who fled to neighboring Chad and CAR. "To date there are no precise figures on the death toll," writes the former governor of Darfur, "nevertheless, reports put the figure of those who have been killed at more than 200,000." (Ateem 2009: 253)

Starting in 1985 the military regime adopted a policy of arming the Arabs of Southern Kordofan to fight the SPLA (a mobilization that continues to this day). The *Janjaweed* were inculcated with an Arab-supremacist ideology, holding that the lineal descendants of the Prophet Mohamed and his Qoreish tribe were entitled to rule Muslim lands, and specifically that the Juhayna Arabs should control the territories from the Nile to Lake Chad. "Promotion of tribal and ethnic conflicts," explains Ateem, "mobilization of the *Janjaweed* militia that continued to torch villages from the mid-1990s through 2002, in addition to long term marginalization of the region, were the major factors that triggered the current political conflict in Darfur." (ibid:. 267) For the past four years a new front of civil war broke out between government forces and two armed rebel movements: the Sudan Liberation Army (SLA) and the Justice and Equality Movement (JEM). The latter famously made an assault on Khartoum in 2008, and continues to menace the regime of Omar al-Bashir.

It was a book entitled *Imbalance of Power and Wealth in Sudan*, popularly known as the "Black Book," prepared and published anonymously in Arabic by former members of the regime who joined the armed resistance in Darfur, that provided the first extensively detailed documents of regional inequalities based on statistical data. The "Black Book" showed how the Center, with 5.4 percent of the population, held 79.5 percent of the executive offices, 70 percent of the ministers, and 67 percent of the attorney generals. Government allocations to the Center accounted for 76 percent of total revenues. (cf. Deng 2009: 46–7) But most important, since its mysterious appearance in 2002, all of the people of Sudan have come to learn about the concentration of political and economic power in the hands of only three Arab tribes, making up just over 5 percent of the total population. "It mentions specifically the Shaigiyya, the Jaalyeen (who consider themselves descendants of the Prophet's uncle, El Jaaly) and the Dongollawis." (ibid.: 45) All three of the successful coups in Sudan were led by members of these three tribes, who also control all the top positions in the state security apparatus. The power of the "Black Book" is that, for the first time, people in the peripheral regions could seek to build alliances not only with one another, but with Sudanese from the supposedly privileged core. It transformed the struggle from a regional conflict between the core and the periphery to a national struggle for liberation from a ruling oligarchy led by three tribes.

Now, you may be asking, where is the *oil* in all this conflict? As should be clear, this is not a resource war, nor should we think about these numerous armed rebellions as being primarily motivated by greed. The reality is that the civil war came first, and then the oil came after. Oil had first been discovered in the early 1970s by the American Chevron Oil Corporation. But when the SPLA attacked Chevron's concession and took company workers hostage, it decided the region was not safe for massive investments (a pipeline would be required to bring the oil online), so Chevron left. Later on in the 1990s, sanctions were imposed by the US government against the regime, prohibiting direct investments in the Sudanese oil sector. Thus the oil reserves remained underground, waiting for the right moment, which came in 1995, when Al-Bashir travelled to Beijing and negotiated an oil deal with the Chinese regime. The Chinese National Petroleum Company (CNPC) entered Sudan in 1996 to become the operator for the Sudanese state Greater Nile Producing Consortium (GNOPC) on the former Chevron concession in the Muglad basin. Three aspects attracted the Chinese: (1) the presence of vast oil fields waiting to be developed, (2) the absence of competition from the major Western multinational oil corporations, and (3) immense potential petroleum resources which could guarantee a constant supply, once the country became politically stable. (Lei 2008: 217)

The CNPC successfully drilled for oil on blocks 1, 2, and 4, discovering the Heglig and Unity fields, where it owned 40 percent of the concession, with its partners Malaysian Petronas (30 percent), the Indian Oil and Natural Gas Corporation (25 percent), and Sudan's state oil firm Sudapet (5 percent). But once oil was discovered, it had to be brought on line, and that meant the construction of a pipeline through some of the most dangerous, war-torn regions of a country in the midst of violent civil war. According to human rights activists, the CNPC subsidiary China Petroleum Technology and Development Corporation (CPTR), which won the bid for construction of the pipeline in the South to Port Sudan in the North, received military protection from the regime, which used Chinese helicopters and machine guns to clear away rebels and civilians living along the pipeline route. The pipeline was completed in

1999, and Sudan became an oil-exporting country, with China as its principal trading partner in oil and arms.

The Chinese later discovered the Fula and Abu Ghabra fields on block 6 in 2001, a concession owned by CNPC (95 percent) and Sudapet (5 percent), with a billion barrels of estimated reserves. This field was connected by pipeline to Khartoum in 2003. The Chinese discovered Palogue and Adar Yel fields on blocks 3 and 7 in 2002–3, owned by CNPC (41 percent), Petronas (40 percent), Sudapet (10 percent) Chinese Sinopec (6 percent) and Kuwait's Tri-Ocean Energy (3 percent), with one billion barrels estimated in the former, and five billion in the latter. By 2006 these new fields were connected by pipeline to Port Sudan. (Morin-Allory 2008: 232–3) As each new oil discovery was brought on line, however, the Chinese created a new structural reality in Sudan. First, the economy transformed into an oil-rentier economy, and the state transformed into an oil-rentier state. (Oil revenues were welcome to the Al-Bashir regime, hard pressed for money and arms to fight the civil war.) Second, the location of the oil fields were mostly in the southern half of the country, but the pipelines carried the production into Port Sudan in the northern half, thereby linking the economies of the North and South.

This had the effect of making the conflict more intense over oil-rich regions, but it also forced the government to negotiate peace in order to protect its new sources of oil revenues. The first peace accord it signed was with the Nuba Mountains Alliance Party in 2002, which started a transition period during which the region would be administered separately (from both the North and the South) under international supervision. At the end of the transition period the Nuba people were supposed to have a referendum in which they would choose: (1) to join Southern Sudan, (2) to join Northern Sudan, or (3) to become an independent state. Meanwhile they would hold representation in the national legislature in Khartoum, and receive a "just distribution of the national wealth." (Salih 2009: 282)

Then in 2005 the government signed a Comprehensive Peace Agreement with the SPLM/A, ushering in a return to the region's autonomy, and promising to hold a referendum on Southern Sudan's independence. Al-Bashir agreed to share power with the SPLM, offering John Garang the position of first vice-president in the government of Sudan and the presidency of the government of Southern Sudan. The day before Garang's inauguration, on January 9, 2005, millions of people flooded into the capital in a historic manifestation of support. "It was clearly a hero's welcome that brought together Sudanese from all parts of the country, and especially the marginalized groups of the South, the Nuba Mountains, Southern Blue Nile, Beja region in the East, and Darfur in the West." (Deng 2009: 49–50) The presence of all of these regional peoples defied the simplistic idea of a North–South divide, not to mention the even more simplistic idea that Garang had been motivated primarily by greed.

After five decades of armed struggle the people of the South had achieved their own liberation, and in so doing, had begun the process of liberating all the peoples from a tyranny. Garang laid the foundations for a democratic, multicultural, secular, unified New Sudan. Then, three weeks later, while returning from talks with the president of Uganda, John Garang died in a helicopter crash on Saturday, July 30, 2005. The news sent shock waves through the country. He was soon replaced by his successor, Salva Kiir Mayardit, deputy chairman of the SPLM, and chief of staff of the SPLA. Kiir had been a general in the armed struggle, but he is more inclined to secession than his predecessor. In 2009 Kiir told his supporters that their choice was between "voting for unity and being a second-class citizen in your own country, or vote for independence and be a free person in an independent country." As the deadline for the 2011 referendum drew nearer, what had been a conflict over identity started to transform into a series of disputes over the country's oil resources. In 2007 violent clashes returned to South Kordofan over control of that oil-rich region. In 2008 a self-proclaimed Baggara Arab even declared the region an independent Arab republic. (Salih 2009: 290) In 2009 JEM rebels from Darfur, where the Chinese are exploring for oil (*Wall Street Journal* 2008) attacked the capital. And the vote for the secession of Southern Sudan has threatened to return the Sudan to a new war.

The Comprehensive Peace Agreement required the SPLM/A to advocate the advantages of unity, and not just those of secession. But John Garang's multicultural vision of a New Sudan appears to have given way to a singular Southern identity that, if copied in Darfur and other peripheral regions of the country, could blow the old Sudan apart into fragments.

9
Popular Resistance and People Power

THINGS DON'T FALL APART

In February 1994 an American journalist named Robert Kaplan published an influential article in the *Atlantic Monthly* entitled "The Coming Anarchy," which argued that population growth, urbanization, and resource depletion were undermining fragile governments across the developing world and represented a threat to the developed world. He had visited West Africa, and his most convincing cases were drawn from that region. His article was hotly debated and widely translated. Thomas Friedman, a columnist for the *New York Times*, called Kaplan one of the "most widely read" authors defining the post-Cold War era, alongside Francis Fukuyama, Samuel Huntington, and Paul Kennedy. Kaplan eventually published a book based on his article (Kaplan 2000), considered a cynical restatement of realism in international relations. A glowing review of this book by the *New York Times* complimented Kaplan for conveying "a historically informed tragic sense in recognizing humankind's tendency toward a kind of slipshod, gooey, utopian and ultimately dangerous optimism." (Bernstein 2000) But his critics accused him of misreading the historical causes of legitimate conflicts as mere "anarchy," and of falsely extrapolating from specific conflicts in West Africa to a general anarchy throughout the developing world. (Bandar 2005)

Ten years later "The Coming Anarchy" has become standard hack term for a frightening menace whenever a domestic-internal conflict defies simplistic explanation. Whether or not people read Kaplan or use his expression, there is a pernicious belief that the complex conflict in the Niger Delta is a harbinger of coming anarchy. This is not the first time that breakdown in social order has been suggested for Nigeria. Chinua Achebe, borrowing from the Irish poet William Butler Yeats, entitled his famous novel about destruction of traditional society in Southern Nigeria by missionaries *Things Fall Apart* (1958). The difference is that for Achebe, anarchy was destruction of precolonial African traditions, whereas for Kaplan, the coming anarchy is the destruction of postcolonial European ones. Readers most frightened by "The Coming Anarchy" were probably those living in affluent countries of the global North. Kaplan seemed to be saying to them that the anarchy evident in the developing world would spill over international borders and invade their peaceful lives. The anarchy is spreading now in Africa, he suggested, but soon it will be coming to you.

Thus there was urgency for something to be done to stop this coming anarchy. Since the conflicts were defined as anarchy, they had to be suppressed. Such false reading of African conflicts has led many readers of the *New York Times* to be against armed militia in the Delta, to treat them as common criminals, or pirates. If left unchallenged, this kind of false reasoning might eventually lead to international intervention in the Niger Delta. In a crusade against a chimera of anarchy, the United States may one day send its troops to help Nigerian soldiers fight piracy, hijacking, sabotage, and disorder. For the United States is the major importer of Nigerian crude oil; and all of that oil comes from the Niger Delta. Now it is clear that such a reading of the conflict in the Niger Delta is false. Scholarly studies on the conflict have demonstrated real grievances by the local populations against both the oil industry and the government (discussed below). But what is more paradoxical is that, while the violence and disorder grows, the Nigerian state remains sovereign, and shows no sign of losing its power. Despite the belief that the present disorder will lead to anarchy, the reality is that things have *not* fallen apart. This raises the question, why not?

Kaplan's celebrated idea was not original. The real theorist behind the coming anarchy was Thomas Homer-Dixon, widely regarded as the central figure in the environment and security debate. Kaplan interviewed Homer-Dixon; then vulgarized his scientific theory, known by scholars as the "Homer-Dixon hypothesis." This theory suggests that the growth of human population and the economy will spur ever-increasing demands for natural resources. This will create growing scarcities of such vital renewable resources as cropland, fresh water, and forests. These environmental scarcities will have profound negative social consequences, such as insurrections, ethnic clashes, urban unrest, and other forms of violence, especially in the developing world. Kapan credited Homer-Dixon explicitly in his work. But Homer-Dixon's scholarly monograph *Environment, Scarcity and Violence* (2000), later published by Princeton University Press, just never enjoyed the same best-seller status of Kaplan's coming anarchy. So supermarket shoppers who bought a copy of Kaplan never really got the whole argument. Homer-Dixon was careful to point out that the effects of environmental scarcity are indirect—they act in combination with other social, political, and economic factors. Capable states, efficient markets, and an educated populace can reduce the likelihood of conflict under conditions of scarcity. Hence scarcity and violence do not necessarily lead to anarchy.

This is an important point of departure for the present chapter, which concerns itself with oil and violence. In the last chapter the conflict in Sudan was related to the presence of oil. But in Sudan oil was not the cause of the violence. The civil war in Sudan came *before* the oil. In this chapter it will be shown how oil has been the cause of the violence in the Niger Delta. For 50 years oil exploitation has polluted the environment so profoundly that its large human population can no longer survive with their traditional rural mode of production. Flaring of gas by the oil companies has raised the temperatures around local communities to levels where people live in a kind of hell on earth. The air is hard to breathe. The water has been contaminated by oil. The fish are dead. Where once the Niger Delta could feed millions of people, fishermen must now enter into other economic activities in order to survive and feed their families. Where once crops could grow, where fruits could be picked from mangrove forests, now villagers must buy their food by selling themselves for money to the highest bidders. Before oil, the Niger Delta was a rich and fertile environment. After oil, the Niger Delta is a wasteland. This industrial "waste" has caused resource scarcities vital to the survival of local populations. Such environmental scarcities alone would be sufficient to explain grievances of local populations, who resort to criminal activities and violent forms of resistance *in order to survive.*

But there is another way in which oil causes violence in the Niger Delta. This has to do with the context of bad governance under a "failed state." Expropriation of the natural hydrocarbon resources by the government, and corrupt enrichment by its kleptocratic elite, mean that local grievances indirectly caused by oil have exploded. Context matters. While the new capital city of Abuja, built on a plateau in the center of the country to house the federal government, has enjoyed billions of dollars of investment in its infrastructure, and swallowed the oil rents in a vortex of greed, poverty and misery have increased in the Delta.

The "oil-rich" Delta remains the poorest and least developed of all the regions of Nigeria. Oil has entered the mentality of Nigerians. It is identified with wealth. It motivates the federal government to collaborate with foreign corporations who pollute air, land, and water. It also motivates the local people to demand compensation. It motivates them to fight, some for justice, and others for greed. It came *before* the war, and without it, there would be no uprising in the Niger Delta. Oil is the *sine qua non* of this conflict.

The Nigerian context has been identified as a failed state, but just what is a "failed state?" The term is often used loosely to refer to a state that fails to meet the needs of its people. But since the people are never satisfied, such a definition would include every state in the world. Rotberg (2004) considers states to have failed when they are "consumed by internal violence and cease delivering positive political goods to their inhabitants" (ibid.: 1), which at least has the merit of distinguishing failed states by their violence. Still, it does not distinguish a failed state from a successful violent tyranny. Before giving a list of definitions, therefore, it must be recognized that any definition of a failed state will depend on what one considers to be a "successful state."

It has become conventional to use the Weberian model as a basis for evaluating state failure. (Weber 1978 [1922]: 56) Stripped to its most basic essentials, a Weberian state is characterized by a single centre of power that holds a legitimate monopoly of violence, controls the population, and rules over the territory where the population lives (internal sovereignty). Also it must be recognized by the international system (external sovereignty). This does not require that it meet the needs of its people, but the problem with using this model is that the postcolonial state in Africa has fundamentally different historical origins, and different present realities, than the European state used by Weber. African states had their borders and institutions imposed by colonial powers. The purpose of colonial rule was extractive rather than developmental. The institutions of rule were not based on legitimate authority, but conquest and raw power. The government rarely controlled its whole territory. In other words, postcolonial states had low degrees of "stateness" to begin with.

More comprehensive criteria for a state have been offered by measures of "governance" which focus less on what a state *is*, and more on what a state *does*. Using such criteria a successful state is one with good governance, and a failed state is one with bad governance. Since these have already been discussed in great detail above (Chapter 3), it will suffice here to restate the criteria of good governance: government effectiveness, regulatory quality, political stability, control of corruption, voice and accountability, rule of law. Annual reporting of governance indicators by international organizations can provide a measure of state failure. One advantage of this approach is that we can speak of state failure or success in terms of degrees. But like the Weberian model, governance indicators measure how much a postcolonial African state resembles the ideal of a Western state, rather than a successful state. My own experience of debating with Chinese scholars about Sudan showed me just how different the criteria of a successful state are in the East and West. The Western model emphasizes the individual rights and freedoms of a liberal democracy. The Asian model of successful states lays greater emphasis on public order and the effectiveness of state power. For us Sudan is failing; for them it is succeeding.

The real importance of the concept of a "failed state" to this chapter is its relationship to what are called "collapsed states." (Zartman 1995) In a collapsed state, central government ceases to function, and is unable to protect its citizens against crime and violence let alone provide for their welfare or maintain the country's infrastructure. There is no single recipe for collapse, nor is there a single set of determinants, but recent history suggests there are some recognizable and recurrent patterns. State collapse is usually preceded by civil conflict of one kind or another, and occurs more easily where the state framework has a relatively thin presence in society, where repression or neglect destroy the regenerative capacities of society, and where people suddenly face deteriorating economic conditions. (Doornbos 2008: 258) All of these signs are present in Nigeria. So the question is not failure of the Nigerian state, but is state failure leading to state collapse?

Few studies of collapsed states in Africa go much beyond the appearance of chaos and crisis. But one scholar, Ricardo Soares de Oliveira, does not see crisis in exclusively pathological ways. He spent five long years with a post-doctoral scholarship at Cambridge working on why empirically failed oil states in the Gulf of Guinea have endured and thrived. His idea of the "paradoxical sustainability of the Gulf of Guinea state" (Soares de Oliveira 2009: 9) observes that oil states in the region, while "failed" because sharing the worst elements of the petro-state and the African postcolonial state, are also "successful" because they are cash-rich, preserved from actual demise, and surprisingly aloof from domestic and external pressures. While the ordinary analysis of state failure in these states implies a prediction that they will collapse, Soares offers an alternative scenario. The state is not collapsing. It is not withering away. The oil state in the Gulf of Guinea remains the central institutional site of the country's engagement with the international economy (petroleum exports) and the ultimate prize in the domestic political struggle (rent-seeking). Far from being irrelevant, on the verge of collapse, or falling apart, the state in a country such as Nigeria finds that its "permanent crisis" may be viable. As the oil state in the Gulf of Guinea is both failed and successful—a logical contradiction— Soares calls this paradox the "successful failed state." (ibid.: 49)

While he agrees that the development of the petroleum industry in Africa has undeniably led to widespread economic deprivation of the majority of the *population*, he follows the reasoning of Reno that corruption and criminality in a failed state flourish to the extent that unparalleled economic opportunity is created for *elites*, who may even have a vested interested in state collapse." (Reno 1995: 29) Oil states produce winners and losers, so whether or not you think your state is failing depends on whether or not you are of the elite. Even individual members of the suffering masses, however, might learn to thrive in a successful failed state by engaging in contraband, plunder, or other illegal activities, such as slavery, narcotics, and piracy. Violent crimes in the Niger Delta are not simply the effects of its environmental degradation. They are also the fruit of a successful failed state: fertile soil for evil men.

How do failed oil states in the Gulf of Guinea survive? Why don't they collapse? When we look at their social indicators— ignoring the high per capita incomes of the smaller countries (a mathematical illusion created by small population denominator and large GDP numerator)—we see that inhabitants of "oil-rich" states are poorer than those of other African nations. On average they have lower life expectancies, lower literacy rates, higher infant mortality rates, and so on. When we look at their indicators of governance, freedom, or perception of corruption, they rank at the bottom of the world. When we look at their security, we find they are more likely to experience violent conflict. In other words, given that every indicator suggests they are failed states, and should be on the verge of collapse, what explains their surprising survival? Soares offers three major explanations: (1) international actors' support, (2) elites discharging their state functions to a bare minimum, and (3) retaining state control over the "useful" parts of the system, namely oil and security.

First, oil companies need states. Oil states are the vehicle that allows access to oilfields. Oil states are the collaborators of the foreign oil corporations. They negotiate contracts, award concessions, and legalize the exploitation of oil. Since the oil companies serve the interests of foreign countries, oil states receive political, economic, and diplomatic support from the international community. This is why oil states lacking domestic legitimacy receive substantial foreign recognition. "The unwavering international need to have sovereign partners underwrites the exploitation of oil" (Soares 2007: 330) and justifies a "realpolitik" agenda. (ibid.: 334) MNCs provide 90 percent of state revenues, and lobby home states for favorable bilateral policies. They are the most important source of "stateness" for African oil states.

Second, since oil rents preclude the need to engage with the population, the elites are able to act without accountability to the people. Not only does this reduce the voice of the people and the accountability of their leaders, it also creates a fundamentally new kind of strategy. During the early decades of independence oil states seeking legitimacy for their rule had pursued grand projects of economic and social development. But over time elites learned that the only legitimacy they needed was from international actors who provided their revenues. Therefore they abandoned the big public works projects and ambitious development plans of the 1960s and 1970s for smaller patron–client networks. These domestic networks have shrunk to include only a minority of the population, "as the major sources of legitimacy reside in the international system rather than at home." (ibid.: 332) This trend was reinforced by the neoliberal agenda of the 1980s, which insisted on fiscal austerity and de-nationalization of the economy. Structural adjustment programs demanded that elites privatize national industries, cut public services, and balance their budgets in order to qualify for debt rescheduling and aid.

During the 1980s and 1990s, an astonishing neglect set in, with many of the white elephant projects falling into decay: "Only a comparatively small portion of the territory—useful spaces such as oil-producing enclaves, elite home-towns and national capitals—is subject to anything resembling an everyday state presence." (ibid.: 330) Thus while they failed the people, the system was largely a success for themselves. In fact, not having to provide public goods was more efficient: "By using the state as a tool of private accumulation but never taking full possession of it, elites can enjoy the prizes of the state without being burdened with its liabilities." (ibid.: 332)

Finally, the enclave nature of the oil industry left oil installations unaffected by the general breakdown of public services in a way ordinary businesses were not. Foremost among the useful spaces are the oil-producing enclaves. Within these walled compounds, or offshore platforms, the primary source of wealth was shielded from the violence and disorder. It is within this geography that the oil states developed their only effective governance. It is within this space that the national oil companies flourished. The national oil companies were essential to keep the downsized state and its oil partnership functioning. While most of the ministries in oil states are empty shells, the national oil companies are the key policy-makers, planners, tax collectors, and negotiators of oil for the government. In many oil states in the Gulf of Guinea, the national oil company is a state-within-a-state. Soares makes this point strongest with the case of Sonangol in Angola, but the idea travels well throughout the region: "As the state withered away, 'the state' retreated into the NOC, the last refuge for educated, technically able personnel working in partnership with international companies and consultants." (ibid.: 96) For similar reasons, the elites in the Gulf of Guinea oil states also maintained their effective security apparatuses. Those oil states with military regimes provided lavish expenditures on the military, disproportionate to actual security threats faced by the country. Civilian regimes spent more on their presidential guards, gendarmes, and secret police, and, following a global trend, hired private security agencies.

Oil states in Africa are failed states but will not collapse, because of international support, short-sighted elite strategies of wealth accumulation, and a cynical use of the state's monopoly of the use of violence. As the preceding chapters have shown, all efforts to help these states stop failing have failed. Foreign direct investment by the multinational oil companies has buttressed the regimes (Chapter 2). International initiatives by non-governmental organizations are not sufficient to improve governance (Chapter 3). Rentier states, military or civilian, are programmed for failure (Chapters 4 and 5). Dissidents are silenced (Chapter 6) and opposition parties behave like the ruling parties they replace (Chapter 7). Perhaps the problem is not how to stop the oil state from failing, but how to collapse the successful failed state.

OIL AND VIOLENCE IN THE NIGER DELTA

Before oil polluted its water, the Niger Delta had vast fertile mangroves teeming with fish. The first oil prospector, a German firm, Nigerian Bitumen Company, arrived in 1908. (Onoh 1983: 42) Fortunately, after Germany's defeat in the First World War, German

Map 9.1   Nigeria

prospecting operations in Nigeria came to an end, and for the next two decades there was no further exploration for oil, initially because of the world recession of 1918–25, then due to the Great Depression of 1929–35. Still, over 90 percent of total investments in Nigeria were under foreign ownership by the time of independence in 1960. (ibid.: 4) In the colonial pattern of foreign direct investment (see Chapter 1), only British companies were allowed to do business in colonial British Nigeria without special permission. So when D'Arcy Petroleum (later renamed Shell BP) began operations in Nigeria in 1937, the Colonial Mineral Ordinance granted the company the entire onshore and offshore exploration and prospecting rights, and a new age of toxic oil pollution began. Although it was given legal rights to all of Nigeria, the firm soon restricted its operations to the delta in the south, where the largest reserves were found.

Colonial development was slow. In 1937 D'Arcy began to search the swampy mangroves. Offshore exploration platforms were not yet developed. There were few roads. Seismic tests were slow and laborious. The company briefly halted its oil prospecting during the Second World War, and resumed operations in partnership with Royal Dutch Shell and British Petroleum in 1946. With additional capital and technology, the first exploration well was drilled in 1951, and a discovery (though not commercial in size) was made in 1953. But it was not until 1956—almost two decades after it had started prospecting—that Shell made its first commercial discovery at Olobiri. It moved its headquarters to Port Harcourt, where there was a deepwater port large enough for tankers, and constructed a pipeline in 1958 connecting its wells to Port Harcourt. By 1959, on the eve of Nigerian independence, Shell had constructed the Bonny oil terminal, from which the world-famous "Bonny light" crude first entered the world oil market to fill the tanks of foreign vehicles around the world.

In the pattern experienced elsewhere in Africa, foreign direct investment changed from colonial spheres of influence to multinational enclaves (see Chapter 1). Pressures from the United States government after the Second World War pushed open the doors to the US oil giants Mobil (1955), Chevron (1961), and Texaco (1963). The French oil company Elf and the Italian oil firm Agip later joined them. These large integrated multinationals remain the major players in the Nigerian oil industry today, accounting for more than 90 percent of the country's crude oil production. (Ukiwo 2008: 76) Niger Delta oil is still dominated by foreign firms. Foreign domination did not change with the arrival of independence in 1960, nor did oil's export orientation. Despite crude production increasing from one million barrels per *year* in 1958 to one million barrels per *day* in 1970 (Onoh 1983: 45), and despite three refineries being built to meet local consumption, almost all of Nigeria's oil was exported as "crude" (97 percent in 1958, and 96.8 percent in 1970). Instead of diversifying into downstream oil industries and using oil for development, Nigeria became a rentier state (see Chapter 4).

As crude oil production increased, associated gas production also increased. However, more than 90 percent of this associated gas was "flared" into the air. Low demand for natural gas was attributed mainly to the climatic conditions of Nigeria. Nigeria lies in the tropics and enjoys an abundance of sunlight, and therefore it requires no gas for heating purposes. Nigerians living in rural areas continued to use firewood for cooking purposes, reasoned the companies, and therefore would not discard this free source of energy for an expensive modern one. Instead of benefiting the Nigerians as a new energy source, natural gas was wasted. By 1981, decades after being discovered, only 46 million cubic meters of total gas was used as fuel, while more than 836 million m³ was flared into the air by firms. (Onoh 1983: 55) Not only was this a waste of a non-renewable resource, but it also had environmental effects on the local populations. Around the flares the air was hot and hard to breathe, and, over a long period of respiration, was carcinogenic.

The first association between oil and violence in Nigeria was the armed secession of the Republic of Biafra and the outbreak of the Nigerian Civil War (1967–70). There is little doubt that the experience of that war was instrumental in heralding the demise of the foreign concessionaire phase. (Adejumboi and Aderemi, in Osaghae et al. 2002) Lagos and Enugu (the capital of Biafra) contested the rights to the oil. Some suspected that the French support for Biafra was a calculated part of the scramble for the oil in the region. After the victorious federal government had suppressed the Igbo rebellion, Lagos embarked on a struggle to "nationalize" its oil industry. This next struggle to wrest control from foreign corporations was legitimized by a postwar nationality discourse: i.e. a conflict between Nigerian and foreign capitalists. This second conflict influenced the trend towards greater state involvement and attempts at nationalizing the oil industry. The underlying idea was to dilute foreign control through increased participation by Nigerian capital, and also to nationalize the very identity of oil. National-ization meant that "Bonny light sweet crude" would thereafter be called "Nigerian" oil. The creation of a Nigerian National Petroleum Corporation (NNPC) promised that all Nigerians would be entitled to production revenues, to promote *national* unity after the civil war (Ukiwo 2008: 78). As the civil war had demonstrated, Nigeria was a state without a nation. Oil was to serve as its main instrument of nation-building.

A new petroleum decree in 1969 vested the entire ownership and control of all oil found under or upon any Nigerian lands, territorial waters, or within Nigeria's continental shelf, with the federal government or its agencies. Only the federal minister of petroleum was entitled by law to grant licenses for exploration, prospecting, or mining. All concessions were to be inspected by federal government-appointed petroleum engineers. All foreign firms were required to recruit Nigerian manpower, to provide infrastructure and utilities in the areas where they operated, and to pay rents and royalties to the federal government. (Onoh 1983: 19–23) Nigeria's admission into OPEC in 1971 encouraged a more aggressive nationalization policy. OPEC member states were obliged to acquire 51 percent equity interests in their oil industries. The new national petroleum company, founded in 1971, had the right to market the federal government's share of crude oil directly to international consumers, not only through the multinationals. It was to participate directly in all aspects of oil operations, recruit and train indigenous manpower, take over concessions abandoned by the foreign firms, and encourage Nigerians to get involved in oil-services businesses such as catering, road construction, laying pipes, and so on. By 1979 the NNPC had acquired 60 percent of Elf, 60 percent of Agip/Phillips, 60 percent of Shell BP, 60 percent of Gulf, 60 percent of Mobil, 60 percent of Texaco, and 60 percent of Pan Ocean fields throughout the country. (ibid: 25)

But Nigeria lacked the capacity, the technical know-how, and the financial reserves needed to take control of its oil industry. So it relied instead on production-sharing contracts that were designed to accommodate new companies in already existing concessions run by Shell, now ostensibly owned by the NNPC. These production-sharing contracts allowed foreign partners to invest capital and to share crude oil *in specie* in proportion to their equity. The foreign oil companies therefore became contractors, and the NNPC a mere rent-collector. Often during the rapid implementation of the program, foreign firms would find a silent partner, someone who knew and cared little about the oil business, but who created the token appearance of a change to national ownership by a visible Nigerian partner. This process was called "fronting" because the visible Nigerian partner would take orders in the front room while the foreign partner or former owner would manage the firm from the back room. A very small number of Nigerians purchased most of the shares of the 952 enterprises nationalized by the first decree. Less than one-tenth of one percent of the Nigerian public benefited directly from nationalization. The Adeosun Commission, a panel later assembled by General Muhammed in 1975 to investigate the affair, estimated that at most 932 individuals acquired shares in half the companies nationalized, and as few as twelve individuals or families had acquired the bulk of the enterprises affected. Such men, mostly southerners, were called "Mister Forty Percent."

One foreign businessman, long resident in Lagos, described the major beneficiaries of the first program as "the people who stole everything from the foreign companies and called it 'indigenization'." (Biersteeker 1987: 97) Yet the actual situation soon evolved differently. By the end of 1982 all oil multinationals in Nigeria operated from a minority equity position, but they retained effective control over their operations. They still had a significant advantage in the exploration process. A drop in oil prices after 1982 allowed them to renegotiate the terms of their production-sharing contracts. Moreover, NNPC held only equity shares, or a share of the profits and production. The NNPC was largely a silent partner, a rent-collector, totally out of the management of the foreign concessions. (ibid: 241) One major national-level political consequence of the nationalization program was its contribution to the growth of state involvement in the economy, which facilitated the rise of a new rentier class, an accumulating bourgeoisie who monopolized all the country's oil revenues for themselves. Nationalization was not a program of national equality.

Challenges to the concept of "Nigerian" oil came from the aggrieved peoples of the oil-rich Niger Delta. Some preferred a designation such as "Bonny light," which specified the place from which the oil was being exploited. They argued that indigenous people of the oil-producing communities should be given privileged access to the rent derived from their natural resources. They saw the "nationalization" discourse as a hypocritical disguise for internal colonialism, and developed a critical "indigenization" discourse that challenged the foundations of Nigerian federalism. (Ukiwo 2008, Naanen 1995) Indigenization strategies took many forms. Sometimes the local peoples living around the oil facilities approached the multinationals for community projects. Other times they attempted to stop production by occupying flow stations or blocking roads. Sometimes local elites demanded that jobs be reserved for them. Others agitated for a larger share of the oil revenues. At its most extreme, indigenization called for exclusion of non-indigenous peoples from occupying posts, claiming that local jobs should be reserved exclusively for local people. (Ukiwo 2008: 82) Perhaps the most famous indigenous-rights activist was Ken Saro-Wiwa, executed by the military regime for speaking this discourse on behalf of Ogoniland. After his execution was revealed, he became an international symbol for the rights of all indigenous peoples in the Niger Delta.

In 1990 a local Shell employee was informed that there was going to be an Ogoni protest against the abuses of the company, which included several thousand oil spills in the region. Shell called the state police commissioner to warn him about the impending protest, and demanded the police provide protection. The police responded to this request by arriving, armed, and shooting 80 Ogoni villagers dead. This was the opening of the struggle for emancipation of Ogoniland from the foreign oil industry and its domestic collaborators.

Ken Saro Wiwa wrote a book entitled *On a Darkening Plain*, which described how the oil companies, in particular Shell, had turned the Niger Delta into an "ecological disaster" and "inhumanized" its inhabitants. At this point in time, few people in the outside world knew how bad pollution had become in the oil region. The international press paid little attention to the grievances of the Niger Delta people, and was at any rate more concerned with the broader abuses of the military regime. But Saro-Wiwa managed to get international media attention, and published the now famous "Ogoni Bill of Rights," showing the effectiveness of dissident intellectuals (see Chapter 6). In January 1993 he rallied 300,000 people, nearly two-thirds of all the Ogoni population, to participate in a manifestation he called "Ogoni Day." He gave a speech in which he declared Shell *persona non grata* and urged all the minorities in the Niger Delta to "rise up now and fight for your rights." (Shaxson 2008: 198) The rally was an immense success, on both the spiritual and material fronts, and resulted in extended protests against the oil company that forced Shell to shut down its operations. This stopped 30,000 barrels a day from flowing to foreign consumers, who now understood there was a problem.

The reaction was predictable, and emblematic of how military regimes and multinational oil companies collaborate in violent repression in Africa. A memo written by the commander of the Internal Security Task Force (a secret police created to suppress dissidents) explained that: "Shell operations still impossible unless ruthless military operations are carried out" [*sic*] and recommended "wasting operations." (ibid.) In May 1994 this taskforce went on a bloody rampage across Ogoniland, killing four Ogoni chiefs, and at least 50 other civilians. It arrested Saro-Wiwa and 15 Ogoni activists, and held them without any access to their lawyers, charging *them* with having killed the Ogoni chiefs! Saro-Wiwa was convicted in November 1995, despite pleadings by Nelson Mandela and others, and was publicly hanged, with eight other Ogoni activists, in a warning to anyone who dared to challenge the right of foreigners to exploit and pollute their lands in collaboration with the military regime.

Shell in fact colluded with the military in this mock trial by bribing witnesses to give false testimony against the Ogoni activists. We know this because, 15 years later, a civil action was brought against Shell in a Manhattan court that charged the company with complicity in the execution of Saro-Wiwa. Wishing to avoid more bad publicity, Shell agreed to pay $15 million to Saro-Wiwa's son and other relatives of the executed activists, a portion of which went into a trust for social programs in the region affected by Shell's oil spills and gas flaring. But Shell's settlement did not provide an admission of guilt. It avoided a trial in which its collusion and pollution would have been aired in court for the whole world to see. Besides, $15 million was like lunch money to a multinational.

Separate from these regional grievances, environmental and social, which manifested in the Niger Delta, the oil curse was an important cause of grand corruption in the federal capital. Many people believe that billions of dollars of oil revenues must somehow trickle down to the general population, and imagine that oil helps poor African countries to achieve development. But since 1970, one percent of the population has accumulated 85 percent of the oil revenue, somewhere between $100 and $400 million of the revenues has gone "missing," while the number of Nigerians living on less than a dollar a day has actually *increased* from 38 percent to more than 70 percent. (Watts 2008: 62) The rise of a rentier class in the federal capital ultimately undermined the goals of nation-alization because the objective of the national leadership was not development. Nigerianization and nationalization were decoys for capital accumulation. (Turner 1980)

Hardly two years after the NNPC was created, the first of many corruption scandals erupted. An investigation revealed that 2.8 billion Naira was missing from NNPC accounts with the Midland Bank of London. President Alhaji Shagari set up a tribunal that exposed secret oil contracts, improper conduct of third-party buyers who banked the proceeds of sales in their own private accounts instead of paying them to NNPC, and customs officials who signed oil loading forms on quantities of oil exported without cross-checking them. (Onoh 1983: 36–7) Such grand corruption has become a regular feature of Nigerian politics, with the country usually ranking at the bottom of Transparency Inter-national's Corruption Perception Index. But during the long dark decades of military rule, grand corruption exclusively benefited an ethno-military class of Northerners who orchestrated the precipitous decline of derivation revenues paid to the states from 50 percent before military rule (1964) to 30 percent under Gowon (1970) to 25 percent under Aboyade (1977) to 5 percent under Shagari (1981) to an all-time low of 1.5 percent under Buhari (1984) and 3 percent under Babangida (1992). (Ukiwo 2008: 80–1) The result is that while the indigenous minorities of the Delta lived in a violent wasteland, and the majority of the population in the country lived in abject poverty, a small class of rentiers lived a pirate's life, and buried their treasure offshore.

COLLAPSING THE FAILED STATE

The new dissident who came to symbolize the struggle of all Nigerians against this venal tyranny was Fela Kuti, a musician who taunted the military kleptocrats with songs such as "Coffin for Head of State." He suffered endless court appearances, police beatings, and torture. During the 1970s, the decade when petrodollars flooded into the treasury, he led a successful form of cultural-political protest on behalf of ordinary Nigerians who were facing a vicious and predatory state. He would strut onstage taunting Nigeria's elites with songs such as "Gentleman" that mocked those who would put on ties, coats, and hats and end up sweating all over, and smelling "like shit." (Shaxson 2008:19) When the military regime splurged over $100 million dollars in unearned oil revenues on an international jamboree called Festac in 1977, Fela organized a Counter-Festac at his music hall Kalakuta. This counter-event was so popular that the military government became enraged. It was already offended by his hit song "Zombie", which portrayed soldiers as mindless robots who would not even think unless they were told to do so. Hundreds of soldiers descended on Fela's compound. "They smashed testicles with rifle butts, they dragged women naked to army barracks and tortured them with bottles, and burned Kalakuta down. Fela was hauled out by his genitals and suffered a cracked skull. His brother Beko was in a wheelchair for months. They threw his 78-year-old mother through a window." (ibid.: 19–20)

Putting Fela in prison, however, only made him a national hero, and when he got out, months later, he continued his dissidence. Fela touched the heart of ordinary Nigerians with his song about military rulers, "Beasts of No Nation," which he wrote in jail during one of several incarcerations. Fela was a dissident voice from another generation, who, alongside other famous Nigerian intellectuals such as Chinua Achebe and Wole Soyinka, had a national stature that no Nigerian dissent enjoys today. They spoke for all Nigerians and provided a "national" voice to counteract the deception of their leaders. But Fela died in 1997, and with him died a certain form of national dissident. "Nigerians have become so polarized by their endless competition for resources, fragmenting the very issues themselves, that it is hard for anyone now to speak for the whole nation." (ibid.: 25) So the loudest dissident voices now come from armed militants who attack oil rigs in the name of their ethnic heroes, against the very idea of a united Nigeria.

In 1999 Nigerian soldiers used helicopters from a Chevron compound and flew them into the Delta state, where Ijaw activists had followed the Ogoni on the path of insurrection. The helicopter squadron attacked two small communities, killed several Ijaw villagers, and wounded many others. Later soldiers and police went to a flow station run by Italy's Agip and killed nine unarmed youths with machine guns. This was the first Ijaw massacre. The youths had been protesting the failure of the company to respect its promises to build some community projects. The oil companies had spent much of the decade trying to buy peace from disgruntled locals, with little success. The companies learned, much to their surprise, that money cannot buy everything. Giving money to one group or community resulted in other groups and communities demanding that they too should receive money, or jobs, or projects. When companies refused to pay new groups, their installations were sabotaged. Since this indirectly threatened the revenues of the state, soldiers attacked their own citizens to defend the interests of foreigners. In November 1999 twelve policemen were killed by an armed gang in Odi Town in the oil-rich Bayelsa state. President Obesanjo, a former army general who had converted to the clothes of a civilian politician, called in the security forces which killed perhaps two thousand people. (Human Rights Watch 2003) This was the second Ijaw massacre.

Today at least 50 separate militia groups exist in the Niger Delta, and the police admit they do not have the firepower to defeat them. It would be a mistake to consider this mass insurrection as an organized politico-military movement fighting for regional secession. These groups have no common leader, and many of them have no official name, identity, or political platform. With the exception of the Movement for the Emancipation of the Niger Delta (MEND), few of these groups have any ideology or recognizable political platform. Many of these armed groups are nothing more than gangs. It would also be a mistake to define this as a war between the government and the militia. Some of these militia work with corrupt state government officials, who pay poor unemployed youths to bunker oil, which is then carried by barge to oil tankers offshore, with the obvious complicity of the state authorities. There are generational, gender, class, communal, and ethnic differences that have generated different kinds of conflicts in the Delta. The sheer diversity of groups and interests has constrained social mobilization towards the realization of common programs. Youths accuse their elders of corruptly enriching themselves. Vandalization of pipelines, kidnapping of expatriates, bunkering of oil straddles the line between political and criminal behavior. Most of all there is no common political movement of the Ogoni, Ijaw, Urhobo, Ogbia, and Itsekiri. Despite its name, there is no indication of a pan-ethnic movement for the emancipation of the Niger Delta. Like street gangs in a ghetto, these groups share a common problematic, but no common organization.

As this chapter has briefly outlined, oil and violence have a long history in the Delta. What we see today however is not a second civil war, nor a war of secession, nor a rebellion, but rather an oil insurrection. Civil war is between geographical sections of the same nation. Secession is withdrawal of a state from a federal union. Rebellion is an organized, armed, open resistance to the authority or government in power. But insurrection is a rising up of individuals to prevent the execution of the law. In the case of the Niger Delta, the goal of MEND is to prevent the execution of the petroleum laws, and to shut down oil production in the region. The Nigerian state is too strong for them to overthrow, yet too weak to eradicate them, and too corrupt to fight their crimes. Nigeria is a "successful failed state" which maintains its sovereignty, but cannot improve the lives of its people. These insurgents are taking matters into their own hands, and if they succeed, will collapse the state which makes their permanent crisis endure.

10
Unscrambling the Scramble for African Oil

"Some things happen of necessity, others by chance, and others through our own agency"

Epicurus (311–271 BC)

Those who diagnose pathologies should also offer solutions. If theories of the paradox of plenty, violent oil conflicts, rentier states, and other problems caused by oil are to be useful, then they ought to provide a blueprint for human action. Theory must be a guide for praxis. Therefore it is advantageous to start with a philosophical idea: The oil curse is not necessary. "The most serious objection to the resource curse theory derives from the empirical finding that the curse does not inevitably materialize but is a mere probability." (Basedau 2005: 325) We have cases where a country depends on oil but does not exhibit any of the symptoms of the oil curse. For Norway, oil has been a blessing, not a curse. It appears that the exact causal mechanisms vary from place to place, and seem to depend on a complex set of other factors, like the regional context, the level of development, the political system, and even on individual choice. "The natural resource curse is not destiny." (Bannon and Collier 2003: 11) It does not *have to* happen.

Of course, some features of the problem are "necessary." History, for example, has left a legacy which cannot be erased. What has happened has happened. We can't unscramble eggs. Geography also structures reality. The oil is where the oil is. Nothing can be done about that. Geology structures scarcity, too. There is a limited amount of oil. When it's gone it's gone. There is nothing to be done about that. Human nature is also full of needs, and humans will seek to satisfy them. There is nothing to be done about that, either. So any real search for solutions to the oil curse must begin with a clear understanding of what cannot be changed. Certain African countries, which have suffered five centuries of historical exploitation, today find themselves geographically located above vast reserves of oil, and they act in ways that are governed by human necessity.

Yet some features of the problem have been caused by "chance," or what we cannot explain. Chance is not a god, or a supernatural force. Chance is a way of expressing our uncertainty. History, for example, contains many recurrent patterns, which we extract through reasoning by analogy, and call "lessons." But we cannot explain everything that happened in the past. History cannot be reduced to logic. Some things happened that defy a neat explanation, such as the causes of complex events with many factors, or where we have rival hypotheses, or when we lack sufficient evidence to make certain inferences. Why were Africans enslaved by Europeans, and not the other way around? Why were some Africans colonized by certain European powers, and not by others? How did the United States and China come to be the dominant players in the African oil industry? Of course, we might offer explanations for any of these. But we cannot call them necessary outcomes. For things could have happened differently. *A fortiori*, theories about the future leave us having to make statements of probability about the likelihood of this or that outcome. How much oil can be extracted? What will its price be in the future? Will oil lead to conflict? Will it lead to development? Answers to these questions depend on so many variables that we must always include chance in our thinking. Otherwise our predictions could be wrong, and our policies misguided.

Nevertheless, some features of the problem are due to "agency." They are the effect of human choices. This is where we can do something.

SOLUTION 1: CONTROLLING CORRUPTION

While human nature is governed by necessity, and evidence of corruption is present in all places and in all historical times, corruption is not necessary. It can be controlled. There is a pernicious argument that corruption is part of human nature, and nothing can be done to eliminate it from the hearts of men. Even if this were true, it does not mean nothing can be done to control corruption. If the ultimate source of corruption is human nature, it is still possible to control corrupt behavior, and to limit its effects. Corruption can be controlled through changing the selection of agents, changing the system of rewards and punishments, gathering information for the analysis of risk of corruption, and by restructuring the relations between agents and users to reduce the monopoly and discretionary power of agents. None of these requires changing basic human nature. Rather they involve changing social institutions. This approach is called "new institutionalism" and it has become an effective tool of anti-corruption activists around the world.

Robert Klitgaard (1988) provides a formula for corruption that conceptualizes the problem in a way that is susceptible to solution. Illicit behavior flourishes when agents have an exclusive power over users (such as a monopoly over a good or service), when they have large discretionary powers (to provide or not provide that good or service), and when their responsibility to their superiors is limited. This was schematically represented by Klitgaard with the following conceptual formula:

Corruption = Monopoly + Discretionary Power – Responsibility

Klitgaard has spent years as a consultant for both governments and non-governmental organizations advising them on how to fight corruption. He provides numerous concrete policies that tackle the three institutional elements conducive to illicit behavior. His first solution involves better selection of agents (i.e. public servants). Usually it is recommended that their selection be based on merit and not on patronage, where patrons reward their clients with public office. However, merit-based selection is not sufficient to prevent corruption. Agents must be selected not only for their ability, but for their honesty. He recommends the elimination of dishonest agents by using precedents of corrupt behavior to disqualify candidates, and by administering psychological honesty tests such as the Reid Report and Trustworthiness Attitude Scale, which is used by businesses to screen out potentially corrupt employees. (Cherrington et al. 1981) Another method he recommends is to use external guarantees of reliability, such as hiring foreign expert accountants, who have both technical competency and a track record in honesty and reliability. (Klitgaard 1988: 80–1)

Klitgaard also recommends changing compensations of agents. For example, increasing salaries of civil servants can reduce the need for corruption. The idea here is not just to raise salaries of all agents, but rather to pay compensations for acts specifically linked to honesty, such as compensating specific agents who fight corruption. Another policy is to draft conditional contracts that link the pensions of agents to their success or failure, giving them long-term incentives that increase over time (the closer they get to retirement). Then there are non-financial rewards such as changing posts, offering work voyages, publicizing honest deeds, providing training, giving praise, and awarding medals (to the most "honest"). This last suggestion is particularly suitable for governments with limited budgets. (ibid.: 83–4)

In addition to the carrot, one can use the stick, by raising the overall level of penalties for corruption, and increasing the punitive powers of superiors. Usually a public servant is legally responsible for his corrupt acts. However, often the only legal punishment is draconian: The agent is sentenced to prison, or loses his or her employment. Not surprisingly, unwilling to enforce these punishments, superiors often renounce punishing subordinates for relatively small acts of illicit behavior. So it is important for them to possess a range of punishments proportional to the wrongs committed. Fines can be imposed proportional to illicit earnings of a corrupt civil servant and to profits of a corrupt user. This way, even the smallest acts of corruption can be penalized, giving punishment its dissuasive force (changing future behavior and not simply sanctioning past behavior). Another tool is the use of unofficial sanctions, such as changing the post of a corrupt agent, or openly publicizing his or her illicit behavior. These kinds of unofficial sanctions generate social pressure that not only induce agents to voluntarily change their behavior, but can also lead progressively to their loss of professional standing, and ultimately a quarantine from other agents. (ibid.: 87–8)

Other techniques involve improving information gathering for detecting corrupt behavior. For example, it is possible to use statistical analysis of random samples of data sets to discover "red flags" indicating corrupt behavior. It is also possible to conduct random inspections. Accountants, inspectors, and specialized agents can be hired to conduct these, or they can be conducted by supervisors. Frequently governments create special anti-corruption agencies exclusively dedicated to crime detection. Using information from outside sources such as media, banks, and even users of a public service (creating special complaint "red telephone" numbers) can increase the amount of information available. Finally, it is possible to reverse the burden of proof on agents and users, requiring them to prove their innocence once information of corrupt behavior has been detected. (ibid.: 89–93)

Restructuring the relations between agents, users and chiefs can also control corruption. Where the public agent has a monopoly on a good or service, this creates an institutional context conducive to corruption. One way to eliminate this inducement is to introduce competition from the private sector. Users given a choice between a corrupt public service and a competitive private provider can dry up illicit sources of revenue through free choice. Reducing the discretionary power of agents is also important, for the more discretion a public servant has to provide a good or service, the more inducement there is for illegal payment by users to acquire it. One can reduce the discretionary power of agents by making them work in teams, instead of leaving them alone with their temptations. Public services can be broken down into separate tasks so that no one agent has complete control. Similarly, agents can be rotated from one function or office to another. If their mission is particularly sensitive to corruption, the mission itself can be modified. Associations of users can also be organized, which function as pressure groups on the public service, reporting abuses of power, and sharing information with other users. (ibid.: 93–7)

All of these recommendations focus on domestic reforms. But it is also necessary to make international reforms. In particular there is a desperate need to regulate offshore banking and tax havens. Attacking bank secrecy is essential to fighting corruption. "Well intentioned efforts to tackle the flows of corrupt money often fail," explains Global Witness in a recent exposé, "because it is difficult to gather evidence against venial ruling elites when they are able to use the sovereignty of their captured and corrupted state as a shield." (Global Witness 2009: 22) This report, *The Secret Life of a Shopaholic: How an African Dictator's Playboy Son Went on a Multi-Million Dollar Shopping Spree in the US,* makes several recommendations for international actors to control African corruption from outside the continent.

First, Global Witness recommends visa denial for corrupt rulers. A visa to the United States or Europe is a highly prized possession, and its denial is an effective sanction. "Unlike prosecution, it can be imposed immediately, and does not require proof of guilt." (ibid.: 22–3) Visa denial is also particularly effective because it can be used to target family members of corrupt officials. So long as corrupt rulers are allowed to travel freely, and enjoy the fruits of their ill-gotten gains, a clear message is being sent that corruption is acceptable. Cutting them off from the possession of their pirate treasures not only holds them responsible for their illicit behavior, it also prevents them from enjoying the fruits of their corruption.

Second, Global Witness recommends making banks do their customer "due diligence" properly. There are many anti-corruption laws on the books. The problem is enforcing them. In another report, *Undue Diligence: How Banks Do Business with Corrupt Regimes* (2009), Global Witness explains how government regulators in the United States and Europe are failing to properly inspect banks which accept money from overseas, thereby not ensuring that due diligence is taken to identify sources of funds deposited by its African customers. The standard of due diligence includes filing of "suspicious activity reports" with authorities when suspected money laundering is taking place. It is up to law enforcement officers to act on such reports. But it is up to banks to file them. Banks must also comply with Financial Action Task Force (FATF) standards by identifying a beneficial owner of any company seeking to open an account with them. Banks are also required to monitor suspicious wire transfers, including accurate meaningful "originator information" (which tells where money originally comes from, and not simply the last bank it was transferred from). Global Witness also urges more transparency over "beneficial ownership," through a standard of national registries to be adopted internationally as a mandatory criterion for trusts and other offshore "shell companies:" "This would require incorporators of companies to identify beneficiaries, and prevent misuse of legal arrangements such as trusts for money laundering." (ibid.: 24)

Finally, to help prevent the money from being stolen in the first place, Global Witness calls for requiring oil companies to report their country-by-country payments to regulators, to reveal how much money they pay to foreign countries for oil. Currently the oil companies report only regional figures, blaming the lack of country-specific data on confidentiality clauses in their contracts with the governments of the oil-exporting countries. More than 400 non-governmental organizations have joined the "Publish What You Pay" coalition in over 70 countries pushing for this global transparency initiative. Corruption is a plant that grows best in darkness, and tends to wither when exposed to the light.

SOLUTION 2: DIRECT DISTRIBUTION OF OIL REVENUES

An IMF working paper on Nigeria suggests taking oil revenues from corrupt rulers and giving it directly to the people: "We propose a solution for addressing the resource curse which involves directly distributing the oil revenues to the public," explain the authors of this radically egalitarian idea. Their proposal has been attacked as unrealistic, but as they argue: "Even with all the difficulties that will no doubt plague its actual implementation, our proposal will, at very least, be vastly superior to the status quo." (Sala-i-Martin and Subramanian 2003) One of the two authors, Arvind Subramanian, co-authored another paper in the prestigious journal *Foreign Affairs* that recommended this same solution for Iraq. "The Iraqi people should embed in their constitution an arrangement for the direct distribution of oil revenues to all Iraqi households." (Birdsall and Subramanian 2004)

Direct distribution of oil revenues is also the big solution proposed by Nicholas Shaxson, who thinks such a system will minimize opportunities for corruption and misappropriation, since windfall revenue stays out of the hands of public officials. "Direct distribution of revenues is a very, very powerful idea," he says. "It works for left-wingers, by redistributing money from rich to poor, and it should please right-wingers, too, by taking money away from governments and handing it to private citizens." (Shaxson 2008: 231) Moreover, this direct distribution scheme can also promote democracy. "Want to spread democracy in an oil-rich country? Take the money from politicians and give it to citizens. Political power follows money." (ibid) Clearly, this is a radical solution. For it does more than attack the oil curse. It attacks the very spirit of our times.

One objection made to this proposal is that such revenues would be comparatively small. Acknowledging that in Nigeria, with its large population, direct distribution of oil revenues would amount to around a dollar a day, and even in Equatorial Guinea, with its much smaller population, would amount to only $30 a day, Shaxson argues, "*This is not the main point.* The benefits are elsewhere." (ibid.: 232; original emphasis) Direct distribution of oil revenues to the people can end the division of citizens by oil revenues. It can disrupt the constant splitting of Nigeria into states (a strategy to acquire a share of the oil revenues which is fragmenting the country). Giving oil revenues directly to the people can gut criminal conspiracies, and break corrupt patronage systems. It can reduce oil bunkering and sabotage, and other criminal activities financially induced by the present state of affairs. Direct distribution can also re-enfranchise citizens, and inject revenue into the local economy. Finally, it could be beneficial to the West by increasing our energy security; at least Africans would "hate us less" for our negligent collaboration with, and enrichment of, their corrupt rulers. (ibid.: 233)

Is it unrealistic? "To get politicians to accept reform," Shaxson admits, "you would need gut grass-roots political action, with international support" (ibid.), but he believes that such a program could provide a powerful shared, unifying goal for change that could realistically build a critical popular mass. As for the international support, "the efforts needed to design and implement this would be small compared to the challenge of dealing with the global threats that emanate from today's tormented oil states." (ibid.) To those who counter that implementation of such a plan would be impossible, "with modern technology—fingerprint and iris recognition and the like—direct distribution should get easier, from a practical point of view." (ibid) Saxson compares the policy to other existing mass-distribution systems. "Poor countries routinely achieve countrywide vaccination campaigns, voter registration, and other grassroots schemes, and rich places like Alaska already do distribute oil income directly to their citizens." (ibid)

This last comparison with Alaska is interesting, because it reminds us that direct distribution of oil revenues to citizens is already being implemented. In the 1970s and 1980s two funds were created by Alaska to share its abundant oil wealth with its citizens. First there were budgetary reserve funds, created in 1976, to stabilize the budget by making loans to the state during difficult times, reimbursed during years of budget surpluses. But, second, Alaska created permanent funds that save revenues for future generations from royalties, revenues, taxes, and signature bonuses. In 1982 a portion of these permanent funds were turned into "dividends" that directly distribute a part of the revenues generated by the funds to all the citizens of Alaska. (Yama Nkounga 2004: 187) In 1990 Norway instituted a state petroleum fund that also serves both budgetary stability and a savings account for future generations. But Norway does not directly distribute revenues to the Norwegian people, at least not to its present generations. Perhaps one day it will. It is feasible.

SOLUTION 3: INVEST IN SOCIAL DEVELOPMENT

A less radical solution is to have the state indirectly distribute oil revenues by investing them in social development. African states are developing countries. In order to accomplish the great transformation from agrarian societies to industrial economies, they need what is called a "developmental state." (Leftwich 1995; Woo-Cummings 1999; Chang 2002) This model is essentially the East Asian model, pioneered in Japan after 1870, and especially after the 1920s, which has been replicated in some other countries like Korea and Taiwan. But it has been successfully adopted in Latin America. African oil-exporting countries need such state involvement because the oil sector does not generate economic development by itself. One of the greatest scandals of the African oil boom is that it has produced far more jobs in the United States and Europe than it ever will in Africa. Oil exploration is by nature capital-intensive rather than labor-intensive, meaning that most investment in Africa goes to developing expensive sophistical hardware like floating production and storage facilities, or loading vessels. Offshore oil exploration is an "enclave industry" where little local input is purchased from the local economy and little output is sold to it. MNCs make few local hires. Oil enclaves are gleaming islands of modernity surrounded by a sea of poverty.

We know that revenues generated by oil enclaves can be harnessed for human development. Venezuela has experienced such an effort under the government of President Hugo Chávez. Following his election in 1998, Chávez has restructured and refined the role of the oil sector, revised existing private sector contracts and radically altered patterns and mechanisms of rent distribution. In Chapter 3 we exposed a failed attempt to impose this model on Chad. As a condition of its support for Exxon's project, the World Bank required Chad to impose strict measures on the management of oil rents, requiring the government to invest ten percent of its oil revenues in a "Future Generations Fund," and the other 90 percent in a "Special Petroleum Revenue Account" that would spend oil money on education, health, rural development, infrastructure, and environment. Law 001 of the Republic of Chad (1998) was passed the same year that Hugo Chávez came to power in Venezuela. So the question is: Why did it work in Venezuela, but not in Chad?

South American countries such as Venezuela are different from African countries such as Chad. They are characterized by established and uncontested territorial boundaries. They have coherent nation-states. And they tend to have a state presence across their national territories. "Colonialism, racial intermingling and repression of indigenous cultures also generated a high level of religious, linguistic and cultural homogeneity" that overrode the primacy of ethnic difference, "which was not parlayed into identity-based politics in the later democratic period." (Buxton 2008: 202) The result is that in South America, antagonisms have historically been channeled on class and not ethnic lines. Struggles have been articulated through ideological lenses of socialism, populism and anti-imperialism. "The South American discourse has always been framed (or disguised) as a positive sum game and legitimized through reference to the national interest." (ibid.: 199) Political struggles in South America tend to be defined as a struggle between a poor majority and elite minority (supported by the United States). The history leading to the Bolivarian Revolution in Venezuela, from a nineteenth-century liberal foreign investment regime to a twentieth-century shift toward resource nationalism, "followed from a transition from the oligarchic regimes, to a new party political system that represented an alliance of working- and middle-class interests," which built their popular appeal and revolutionary legitimacy around demands for democratic reform and economic nationalism. (ibid.: 203) In Africa, this third wave of democratization happened only late in the 1990s, and coincided with hegemonic neoliberalism.

Latin America's orientation toward economic nationalism in the 1960s was reinforced by proponents of a different kind of strategy: import substitution. Raul Prebisch's *The Economic Development of Latin America and Its Principal Problems* (1950) proposed to reverse the unfavorable terms of trade of developing countries with developed economies by engaging in manufacturing and production activities for their own domestic markets. Import substitution strategies were designed to re-invest export revenues in the national economy. "The adoption of state capitalist models of development across South America assisted in the legitimization and consolidation of early democratic systems," introducing nationalization projects alongside major programs of land reform, unionization of labor, and rudimentary welfare states. (Buxton 205) Although economic nationalism came relatively late to Venezuela, once adopted it was relatively pluralized and equitable. "In effect, all sectors benefited from the newly enacted distributive capacity of the state, mediated by the dominant party." (ibid.)

Then came Margaret Thatcher and Ronald Reagan in the 1980s, bringing with them a new era of economic liberalism marked by a dramatic shift from closed to more open economies. For the next two decades Latin American countries were pushed to shift from nationally focused development to regional integration strategies underpinned by trade and financial liberalization. They implemented steep reductions in government spending. During the 1980s and 1990s the implementation of these neoliberal strategies deepened levels of inequality, excluded informal workers, the poor, and even middle-class groups. Opportunities for rent accumulation during this period quickly contracted to a small group of elites with personal ties. Meanwhile the large corporatist parties, which had invested export revenues in social development, were replaced by neoliberal politicians, usually economists trained in the US, who privatized national industries, and created a new class of wealthy businessmen. Hostility to privatization and other neoliberal policies of the "Washington consensus" translated into new support for economic nationalists, especially hard leftists who spoke the language of populism, such as Evo Morales of Bolivia, Rafael Correa of Ecuador, and Hugo Chávez of Venezuela. This regional wave of protest against the new rich and the Yankee imperialists is not an isolated national revolution in Venezuela. Nor is the Bolivarian Revolution a totally new program. In many ways it is an old Latin American class struggle by a poor majority fighting against a rich minority, and their American supporters. Finally, Chavez' election came about in a country that had the longest history of procedural democracy in Latin America, a military that was trained in civilian universities and did not constitute a military class like other more typical south American militaries, and after a period in which, having adopted neoliberalism earlier than any other region in the world, the bourgeois democratic party system had been thoroughly discredited.

Venezuela may enjoy huge oil revenues, but during the neoliberal era these had been mopped up by a tiny percentage of the population. After two decades of American-style capitalism, the great majority of the Venezuelans had become permanently poor and hungry. In 1995 the top 10 percent of the population was receiving 50 percent of the national income, 80 percent of the population was earning the "minimum wage or under," and 40 percent of the population in 1996 was living in "critical poverty." (Gott 2005: 173) With oil revenues rivaling those of Saudi Arabia, this just didn't make any sense. These statistics were well-known to Chávez, and he would constantly tell foreign visitors how difficult it was to explain why such a rich country could at the same time be so desperately poor. So Chávez pursued constitutional reform in 1999 and passed a new Hydrocarbons Law (2001) that renationalized petroleum resources and renegotiated contracts with foreign oil companies. The new law mandated a majority of shares in any oil project to be held by the national oil company Petróleos de Venezuela (PDVSA). All of the oil contracts signed during privatization in the 1990s were revised. Taxes and royalties were increased. These changes fundamentally expanded government revenues. His goal was not, like his kleptocratic African counterparts, to build a new presidential palace, or to steal the oil revenues and hide them in secret offshore bank accounts, but to invest oil revenues in social development.

One of the first things Chávez did was to create a Social Fund (FUS) linking together a number of earlier government organizations that used to deal with heath and social welfare. The FUS, together with the new People's Bank, was designed to carry through social politics that were aimed at improving the health and welfare of the poor majority of the population. Its budget derived from the ordinary budgets of the earlier organizations it had gobbled up, and from the Macroeconomic Stabilization Fund (FEM) that channeled oil revenues into government projects. The FUS received almost half of its budget from the FEM, funding schools, hospitals, churches, and the "Plan Bolivar 2000." (Gott 177–9) This last program was one of Chávez's most original ideas. Announced within weeks of his inauguration, this big idea was to mobilize the spare capacity of the armed forces, link it up with local community groups, and send them out into the countryside to help rebuild roads and schools. Venezuela was divided into 25 action zones, in which some 40,000 soldiers and volunteers began working on reconstruction. All of these moves were attacked by the opposition as "populist." But the opposition was comprised of mostly well-to-do and upper-middle-class Venezuelans who had been the primary beneficiaries of the neoliberal era. It also included civil servants and state employees in the oil sector who were used to siphoning off oil revenues at its source. These two privileged classes attempted to overthrow Chávez, in a political struggle that involved protest marches and a strike by oil workers. The defeat of the oil strike and the subsequent renationalization of PDVSA introduced an entirely new and more radical phase in the Bolivarian Revolution.

In the course of 2003, huge sums of oil money were redirected into imaginative new social programs known as "missions," which were gradually established throughout the country. PDVSA contributed billions of dollars per year to these missions, of which there are 17, providing education, health services, housing, and job creation. The missions were established to bypass the lethargic bureaucracy of the state, which had remained largely in the hands of the upper-class opposition. So educational missions were not run by the education ministry, nor were the health missions run by the health ministry, but instead they were run independently. While it has been alleged that the education provided is not up to the quality provided by non-Chavez schools, it is these missions that could be emulated by African countries such as Chad, rather than the World Bank model, which tries to use the corrupt elites running the ministries for social development.

The most immediately important mission was *Misión Barrio Adentro* (Into the Heart of the Shantytown), a medical program staffed by thousands of Cuban doctors. Working in pairs, these doctors went into the poorest barrios and set up health clinics, at first by squatting in people's houses or community centers, then the next year in newly constructed health clinics. These clinics provided local health services seven days a week, 24 hours a day, with medicines provided free by Cuba. Cuban doctors already had a lot of experience working in tropical countries, and they first put particular emphasis on preventative health care. Chávez built 4,400 such community health clinics, providing free primary healthcare for 68 percent of the population. (Buxton 2008: 208) Another mission, *Misión Robinson*, was a literacy campaign designed to teach one million people to read and write, and to use basic arithmetic. Cuba provided televisions, video recorders, reading glasses, and printed materials. *Misión Ribas* enrolled 600,000 students who had dropped out of school into night programs, with a paid stipend, which allowed them to finish their secondary studies in grammar, mathematics, geography and a second language. *Misión Sucre* paid for college preparatory classes, allowing underprivileged students to gain access to the universities.

According to government statistics, the education missions led to the construction of 3,000 primary schools, assisted 1.4 million people to complete high school, and expanded university access to 27,000 students from disadvantaged backgrounds. (ibid.) But Chávez also subsidized the poor through *Misión Merca*, which distributes cheap food to eight million hungry people in 6,000 state supermarkets. Providing cheap food to the urban poor is one of the single most popular missions, because it fights hunger and prevents malnutrition. But Chavez did not stop with the urban poor. Three other missions deal with rural poverty. *Misión Zamora* provides revenues to poor peasants. *Misión Piar* deals with problems of mining communities. *Misión Guaicaipuro* provides funding for Venezuela's indigenous groups. In 1999 Chávez changed the Venezuelan Constitution to include a bill of rights for indigenous peoples, recognizing their native land rights (Art. 119), requiring consultation before exploiting resources on their land (Art. 120), mandating multicultural and bilingual education that respects their cultural identity (Art. 121), protecting their traditional medicine (Art. 122), and their right to retain their own economic practices

(Art. 123), protecting their ancestral intellectual property from patenting (Art. 124), and guaranteeing indigenous representation and participation in the political life (Art. 126). *Misión Identidad* provides voter registration of excluded groups, including registration of foreign residents from poor neighboring countries who have traditionally been denied the rights of citizens. *Misión Vuelvan Caras* provides jobs for youths who have finished the educational missions, with a goal of reducing unemployment among this disadvantaged category.

All of this is to say that Chávez is not just trying to purchase consent with oil rents, but is seeking a fundamental transformation of Venezuela from an unjust rentier state to a just, oil-rich society. When Chavez was elected in 1998, seven out of ten Venezuelans lived in poverty. Ten years later this number had dropped to less than half. When Chavez was elected, the country's Human Development Index (HDI) was 0.69. Ten years later, it had risen to 0.81. (Buxton 209) The upper-class opposition bitterly opposed the new missions during their first year, and dismissed them as "populist." But by the 2004 elections, even opposition politicians were promising to maintain spending on most of these projects if they were to be elected. Of course, it must be acknowledged that there is a danger of a return to clientism. Since the state oil company is the sole benefactor of social development, those who are not loyal to Chavez can be denied development funds. It might be noted that, despite these constitutional provisions, on the ground there have been indigenous protests against oil pollution created by the state oil company PDVSA. Also, Venezuela has been rocked by several corruption scandals. But there is something powerful about the Bolivarian policy of missions which might be transferred to African oil states.

SOLUTION 4: BOYCOTT AFRICAN OIL

The boycott was invented in 1880 in Ireland, and for more than a century has been used by activists to enforce government and corporate social responsibility. In the original boycott a group of Irish tenants refused to do any business whatsoever with an English landowner—Captain Charles Boycott—because of his cruel eviction policies. In the end the peasants triumphed. This is the model of the traditional retail boycott, and has been copied by groups such as Greenpeace in their effort to stop Exxon from burning shale oil in Canada. There are three basic challenges to traditional boycotts.

The first is that they tend to be short-lived, and even if they are successful they offer no continuation beyond the boycott's end. The second is that it is very difficult to create and to maintain momentum around boycotts, especially when there are no measurements of their successes or failures. The third is that boycotts involve some kind of sacrifice on the part of those who are waging them, that is, the cost of pursuing traditional boycotts is born mainly by those who pursue them. Therefore any traditional boycott of African oil would have to overcome all three of these challenges in order to succeed, as Greenpeace's boycott of Exxon has demonstrated.

Fortunately, the traditional retail boycott has been improved by a new and improved formula: the internet-enabled, market-savvy, hedge-fund leveraged "smart boycott." This novel innovation involves creating a new security in a fund with an internet community that trades on the strengths of different boycotts. It uses the tendency of market hedge funds to "sell short" a stock that they believe may be hurt by a boycott on company sales. The smart boycott was invented by two activists, Max Keiser and Stacey Herbert, who had made their fortune on Wall Street during the internet boom of the 1990s (then got out before the market collapsed). They were socially responsible activists who found themselves with the means to do something with their talents. They founded a new fund called the Karmabanque. The way it works is an activist who wants to do something about the irresponsible practices of a company visits the Karmabanque online, and researches that company's vulnerability to a boycott. They do this by using the Karmabanque's "Boycott Profitability Ratio" (BPR), which measures the impact that every dollar not spent with a company would have on its market capitalization. Then, for as little as $500, a boycott portfolio is opened with the bank. The customer then is allowed to choose up to three companies that are being boycotted. They are essentially placing bets on the share value of the company. If the stock goes down, the hedge fund's "short position" goes up in value. So those who would like to sanction Shell, for example, because of its activities in the Niger Delta, could take advantage of declines in the company's share values (like when there is bad news in Nigeria). As the price of the shares goes down, the value of the "short position" goes up.

The Karmabanque "smart boycott" solves the three challenges of the traditional retail boycott. First, since its customers can keep several boycotts in their portfolios, they are able to shift their dissent to another boycott by "selling" one boycott and "buying" another. This allows them to sustain their dissent, indefinitely if they like.

As Keiser and Herbert have argued, this is the very essence of a boycott: expression of dissent. The kind of people who participate in a smart boycott can sustain their activism in a way that a single, isolated, traditional retail boycott rarely allows. Second, since the market data provides them with feedback about the success or failure of a boycott on a company's share value, this creates the conditions for building momentum. Boycotts become, in economic language, "accretive." When they see that one boycott is succeeding (the company share is falling) they can shift their dissent to that successful boycott, increasing its overall effectiveness. When they see that another is succeeding, they can then shift to that one. Often boycotts become successful only when they reach a critical mass. The Karmabanque provides a mechanism that can quickly produce just such a threshold, at which point a company will change its policies in order to protect its market capitalization. Third, unlike a traditional boycott, which places the cost on the shoulders of those who pursue it, the "smart boycott" inversely generates a reward for them. Individuals actually profit from a successful boycott. This has the genius of building off one of the most powerful human motivations: greed. For example, the Karmabanque actually posted the following advertisement: "Attention Arabs: Make Money Boycotting Coca Cola." People who might not ordinarily get involved in a boycott for reasons of social conscience can be induced to participate in a "smart boycott" for purely financial reasons.

Imagine that the price of a barrel of Bonny Light Sweet Crude included the cost of cleaning up the Niger Delta. Greenpeace has tried for decades to "monetize" such environmental harm by ensuring that the whole cost of a product is reflected in its price. Unfortunately for Greenpeace, a traditional boycott of Exxon for its shale oil activities has not impacted the company very much because multinational oil companies are highly insulated from a *retail* boycott. Only a small percentage of a multinational corporation's profits come from consumers at the pump (i.e. distribution). More profits are made upstream.

"The stock price for a commodity should reflect the interests of not only the shareholders, but the stakeholders, and the global community in which it operates," says Keiser. "The real challenge is how to bring an economy of scale to activism." (Greenpeace 2003) In using the indirect forces of the marketplace, the Karmabanque offers its customers, whom they identify as "activists, anarchists, and hedge funds," the same instruments that multinationals have used to increase their capitalization beyond what would ever have been possible were they only to have been in the oil business. But the smart boycott is an indirect tool that depends on the existence of a traditional boycott. It is making a bet on the success of a traditional boycott. Therefore for a smart boycott of African oil to work, there must first be a traditional retail boycott of companies doing business with corrupt African oil regimes. When you buy coffee, or wine, you know where it comes from. You know its provenance. But when a consumer buys gasoline at the pump, no information is ever provided about where it comes from. This is because the oil companies blend different crudes at the refinery, they claim, but it does not seem technically impossible to inform consumers about the blend. At any rate, for the moment, it would be very difficult to boycott Equatorial Guinean oil directly. Therefore a traditional retail boycott would have to target companies doing business with Equatorial Guinea. It would have to ostracize companies that sign exploration and production contracts upstream, or who purchase imports from the country, transport, refine and distribute it downstream. How could this be done?

One approach would be to adopt the successful model of the Kimberly Process, which through its certification procedures managed to boycott "blood diamonds" coming from Angola, Liberia, and Sierra Leone. It is possible to imagine a similar certification process for "blood oil" that boycotted crude oil coming from African countries where oil is the cause of conflict, corruption or dictatorship. Countries that failed to spend their oil revenues on social development could be targeted, as well as those where oil fuels civil war. This idea has been proposed before. But it has its critics. "It is questionable whether one can transfer the relatively successful Kimberly Process mechanism in the diamond sector to the oil sector." (Basedau 2005: 342)

The problem is that diamonds are fundamentally different from oil, because they have a very limited strategic significance for the global economy. It is easier to boycott diamonds, because people do not need diamonds. Also the diamond market is dominated by one player, De Beers, so it was much easier to agree on a common control regime. Oil is not dominated by any single player, and with the emergence of new actors such as China, it is not possible even to imagine a common certification process. Consider the case of Sudan, which would make an ideal "blood oil" campaign, but is supported by China. If De Beers were motivated by wishing to distinguish "blood diamonds" in Angola from its own diamonds that were being produced in Botswana, then the problem that the

West and the East have different standards about what constitutes a good or bad African oil regime remains a serious obstacle to imagining a replication of the Kimberly Process.

SOLUTION 5: STOP CONSUMING OIL

Mahatma Gandhi once said, "We must become the change we wish to see in the world." For those readers who are now fully aware of the many evils expressed by the idea of the oil curse, and who wish to change the problems caused by oil dependency in Africa, there is really no better solution than to reduce, and, when possible, to stop, our own oil consumption. This is individual action that can be taken immediately. It doesn't require sending troops, nor following leaders, nor imposing another international regime. If we cannot change our own consumption, then we should not expect the curse that it causes to disappear.

Until now, our efforts to fight the oil curse at the national and international levels have largely focused on changing the behavior of *other* people. Our governments want African oil regimes to be more democratic. Our corporations want the Africans to sell us their oil. Our international organizations want them to be more transparent and create better governance. Our journalists want them to free their press and allow investigations. Our environmental activists want them to stop polluting. None of these solutions call upon us to change our *own* behavior. All of them have failed.

What this book has argued is that real change has come not from above, but from below. Starting with the failure of major power diplomacy and foreign interventionism (Chapter 1), the irresponsibility of corporate governance (Chapter 2), then the insufficiency of international transparency initiatives (Chapter 3), the book has broken down the growing body of theoretical literature and empirical case studies into separate levels of analysis, and started with an explanation about how and why our international efforts have not effectuated real change in African oil states. So far the world's major powers have acted in ways that serve their own national interests, first through colonialism, then through neocolonialism. As one prominent African succinctly stated the problem, "The end of empire did not basically change these orientations, and the pattern imposed by today's World Bank and the International Monetary Fund is not essentially different from the old colonial system except for two things: mass media cultural seductions have replaced the whip as an inducement to perform, and trade has been multilateral-ized away from the old empire monopolies." (Prunier 2009: xxxiii)

The book further showed why African oil regimes, at the national level of analysis, have been incapable of changing themselves, because massive influx of oil revenues turns them into rentier states (Chapter 4) and because their preponderant military regimes are no better than their civilian ones at channeling oil revenues into social development (Chapter 5). Moving down below the governmental level, to the sub-national level of analysis, it then showed how successes of various kinds of domestic actors, alternatively, have managed to change some aspects of the oil curse. Intellectuals have exposed corrupt rulers through their writings and dissent (Chapter 6). Opposition political parties have used fair and free democratic elections to remove corrupt rulers from power and implement new oil revenue laws (Chapter 7). Armed struggles by subaltern classes have fought for and achieved real autonomy for their regions (Chapter 8) or have successfully shut down polluting oil operations in a kind of environmental self-defense (Chapter 9). For the patient reader who has followed the flow of these chapters in their proper order of presentation, this book arrives by its argumentative structure to the conclusion that real change has come not from above, but from below. All of the positive changes in African oil states have come from some kind of transformation in consciousness that shaped popular resistance to an oil business that is serving *our* interests more than theirs. It is now time for us to experience a transformation of consciousness.

Venezuela may be a model of using oil revenues to promote positive social development. If we could somehow make African rulers more like Hugo Chavez, and African societies more like Latin American societies, then perhaps African oil revenues could be used for positive social change. But those who think they can create such changes through international pressure or military intervention should ask themselves: Was the Bolivarian Revolution the result of our foreign policy? Was it the result of foreigners concerned with poverty and suffering of Venezuelans? Was it the result of our non-governmental organizations? Was it the result of our multinational corporations? On the contrary, it was a radical change that came from Venezuelans themselves.

This is not to suggest that we in the developed countries should do nothing, or that we should only act individually. The truth is that there is only so much that we can do alone. Some kind of collective action is called for. There is a pressing need for people to demand that our governments move quickly from dependence on fossil fuels to renewable energy. International initiatives to establish new mechanisms such as caps on total carbon emissions, carbon taxing, and making polluters pay represent a start of a new kind of citizen politics. Taking away the big subsidies governments pay to the fossil fuel industry, and shifting subsidies to renewable energy research, represent another. An international fund should be set up by the UN to help African countries get off fossil fuels. Perhaps this could be paid for with assessments on rich oil consuming countries, or by a tax on international financial transactions or on arms sales. Examples of successful shift to renewables can be found in Portugal, Spain, and Germany, where governments have found the will to change old carbon habits. But to wait for our governments to act on our behalf, naively counting on wise leaders to betray the interests of large global energy corporations is a grave mistake. "If there's going to be change, real change," said Howard Zinn, "it will have to work its way from the bottom up, from the people themselves. That's how change happens." (Herbert 2010)

Those who really want to help Africans escape their oil curse should focus on getting their own houses in order, and stop rearranging the furniture in African dwellings. We should reduce, and, when possible, stop consuming oil. Some might object that this is not possible. It seems too hard to imagine our world without oil. But the reality is that one day we will run out of oil, and our consumption will stop. Therefore proposing that we should stop now instead of later is more realistic than it at first appears. Moreover, the pollution generated by oil consumption and the dramatic effects it is having on the global environment—air, land and sea—could make the injustices it causes in Africa seem like something of an afterthought. There are many reasons to stop consuming oil. The African oil curse is only one of them.

Bibliography

Adekeye, Adebajo, *The Curse of Berlin: Africa After the Cold War*, New York: Columbia University Press, 2010.

Alier, Abel, *Southern Sudan: Too Many Agreements Dishonoured*, Exeter: Ithaca Press, 1990.

Almond, Gabriel and Sidney Verba, *The Civic Culture: Political Attitudes to Democracy in Five Nations*, Princeton, NJ: Princeton University Press, 1963.

Al-Rahim, Muddathir Abd, *Imperialism and Nationalism in the Sudan: A Study in Constitutional and Political Development 1899–1956*, Oxford: Clarendon Press, 1969.

Amin, Samir, *Unequal Development: An Essay on the Social Formation of Peripheral Capitalism*, New York: Monthly Review Press, 1976.

Amnesty International, *Equatorial Guinea: A Country Subjected to Terror and Harassment* (January 1, 1999).

Ateem, S.M. Eltigani, "Anatomy of the Conflict in Darfur," in Deng 2009: 253–74.

Atenga, Thomas, *Cameroun, Gabon: la presse en sursis,* Bruyères: Editions Muntu, 2007.

Auty, Rick, *Sustaining Development in Mineral Economies: The Resource Curse Thesis*, London: Routledge, 1993.

Ballard, John, "Four Equatorial States," in Gwendolen Carter, ed., *National Unity and Regionalism in Eight African States*, Ithaca, NY: Cornell University Press, 1966.

Bandar, Omar, "A Refutation of Robert D. Kaplan's Thesis," unpublished thesis, American Graduate School of Paris (December 2005).

Bannon, Ian, and Paul Collier, eds, *Natural Resources and Violent Conflict: Options and Actions*, Washington, DC: World Bank, 2003.

Basedau, Matthias, "Resource Politics in Sub-Saharan Africa Beyond the Resource Curse," in Matthias Basedau and Andreas Mehler, eds, *Resource Politics in Sub-Saharan Africa*, Hamburg: Institute for African Affairs, 2005: 325–49.

Bayart, Jean-François, *The State in Africa: The Politics of the Belly*, London: Longman, 1993.

Bayart, Jean-François, Steven Ellis, and Beatrice Hibou, *The Criminalization of the State in Africa*, Bloomington: Indiana University Press, 1999.

Beblawi, Hazem, "The Rentier State in the Arab World," in Hazem Beblawi and Giacomo Luciani (1987): 85–98.

Beblawi, Hazem, and Luciani, Giacomo, eds, *The Rentier State, Vol. 2, Nation, State and Integration in the Arab World*, London: Croom Helm, 1987.

Benda, Julien, *Le Traison des Clercs*, Paris: Grasset, 1927.

Benedetti, M., "Les Iles Espagnoles du Golfe de Guinea: Fernando Poo, Corisco, Annobon," Communication du Ministère des Affaire Etrangères (Direction des Consulates et Affaires Commerciales), *Bulletin de la Société Géographie* (Paris, 1869): 66–81.

Bernstein, Richard, "The Coming Anarchy: Dashing Hopes of Global Harmony," *New York Times* (February 23, 2000): E9.

Beti, Mongo, *Ville Cruelle*, Paris: Présence Africaine, 1954.

— *Le Pauvre Christ de Bomba,* Paris: Robert Laffont, 1956.
— *Mission Terminé,* Paris: Buchet-Chastel, 1957.
— *Le Roi Miraculé,* Paris: Buchet-Chastel, 1958.
— *Main Basse sur le Cameroun,* Paris: François Maspero, 1972.
— *Perpétue et l'Habitude du Malheur,* Paris: Buchet-Chastel, 1974.
— *Remember Ruben,* Paris: L'Harmattan, 1974.
— *La Ruine Presque Cocasse d'un Polichinelle,*Paris: Editions des Peuples Noirs, 1979.
— *Les Deux Mères de Guillaume Ismaël Dzewatama,* Paris: Buchet-Chastel, 1983.
— *La Revanche de Guillaume Ismaël Dzewatama,* Paris: Buchet-Chastel, 1984.
— *Lettre Ouverte aux Camerounais ou la Deuxième More de Ruben Um Nyobé,* Paris: Editions des Peuples Noirs, 1986.
— *Dictionnaire de la Négritude,* Paris: L'Harmattan, 1989.
— *La France Contre l'Afrique,* Paris: La Découverte, 1993.
— *L'Histoire du Fou,* Paris: Julliard, 1994.
— *Trop de Soleil Tue l'Amour,* Paris: Julliard, 1999.
— *Branle-bas en Noir et Blanc,* Paris: Julliard, 2000.
— *Le Rebelle I,* Paris: Editions Gallimard, 2007.
— *Le Rebelle II,* Paris: Editions Gallimard, 2007.
— *Le Rebelle III,* Paris: Editions Gallimard, 2008.
Bickerton, James, and Alain Gagnon, "Regions," in Daniele Caramani, ed., *Comparative Politics,* Oxford: Oxford University Press, 2009: 367–91.
Biersteker, Thomas J., *Multinationals, the State, and Control of the Nigerian Economy,* Princeton, NJ: Princeton University Press, 1987.
Birdsall, Nancy, and Arvind Subramanian, "Saving Iraq from its Oil," *Foreign Affairs* (July/August 2004).
BP, *Statistical Review of World Energy,* 2008, www.bp.com/statisticalreview.
Buijtenhuijs, Robert, "Chad: The Narrow Escape of an African State, 1965–1987," in Donal B. Cruise O'Brien, John Dunn, and Richard Rathbone, *Contemporary West African States,* Cambridge: Cambridge University Press, 1989: 49–58.
Buxton, Julia, "Extractive Resources and the Rentier Space: A South American Perspective," in Omeje 2008: 199–212.
Campodócino, Humberto, *Renta petrolera y minera en países seleccionados de América Latina,* Santiago, Chile: UN Comisión Económica para América Latina y el Caribe, 2008.
Cardoso, Fernando Henrique, and Enzo Faletto, *Dependency and Development in Latin America,* Berkeley: University of California Press, 1978.
Central Intelligence Agency (CIA), www.cia.gov/library/publications/the-world-factbook 'Country Profiles' for Angola, Cameroon, Chad, Congo, Equatorial Guinea, Gabon, Mauritania, Nigeria, São Tomé and Sudan.
Challaye, Félicien, *Un Livre noire du colonialisme: Souvenirs sur la colonisation,* Paris: Les Nuits Rouges, 2003.
Chang, H.-J., *Kicking Away the Ladder,* London: Anthem Press, 2002.
Cherrington, D.J., "The Role of Management in Reducing Fraud," *Financial Executive* No. 49 (March 1981): 23–34.
Clark, John, "Congo: Transition and the Struggle to Consolidate," in John Clark and David Gardinier, eds, *Political Reform in Francophone Africa,* Boulder, CO: Westview Press, 1997.
Collier, Paul, "On Economic Causes of Civil War," *Oxford Economic Papers,* 50 (1998): 563–73.

— "Doing Well Out of Civil War: An Economic Perspective," in Mats Berdal and David Malone, eds, *Greed and Grievance: Economic Agendas in Civil Wars,* Boulder, CO: Lynne Rienner, 2000.

— *The Bottom Billion,* London: Oxford University Press, 2007.

Collier, Paul, and Hoeffler, Anke, *Greed and Grievance in Civil War,* Washington, DC: World Bank, 2001.

Collins, Robert, *Shadow in the Grass: Britain in the Southern Sudan 1918–1956,* New Haven, CT: Yale University Press, 1983.

Coquery-Vidrovich, Catherine, *Le Congo au Temps des Grandes Compagnies Concessionaires, 1898–1930,* 2 vols, Paris: EHESS, 2002 [1972].

Copinschi, Philippe, "Stratégie des Acteurs sur la Scène Pétrolière Africaine (Golfe du Guinée)," *Revue de l'Énergie,* No. 523 (January 2001): 18–44.

Dahl, Robert, *Polyarchy,* New Haven, CT: Yale University Press, 1971.

— *Democracy and Its Critics,* New Haven, CT: Yale University Press, 1989.

— "A Democratic Paradox?" *Political Science Quarterly,* 115/1 (2000): 35–40.

Daly, M.W., *Empire on the Nile: The Anglo-Egyptian Sudan: 1898–1934,* Cambridge: Cambridge University Press, 1986.

Davidson, Basil, *In the Eye of the Storm: Angola's People,* rev. edn, New York: Penguin, 1975.

Decalo, Samuel, *Coups and Army Rule in Africa: Studies in Military Style,* New Haven, CT: Yale University Press, 1976.

Deffeyes, Kenneth S., *Hubbert's Peak: The Impending World Oil Shortage,* Princeton, NJ: Princeton University Press, 2001.

DeLancey, Mark, and M.D. DeLancey, *Historical Dictionary of the Republic of Cameroon,* 3rd edn (Metuchen, NJ: Scarecrow Press, 2000).

Deng, Francis, *New Sudan in the Making,* Trenton, NJ/Asmara: Africa World Press, 2009.

Deroo, Eric, and A. Champeaux, *La Force noire: gloire et infortunes d'une legend colonial,* Paris: Editions Tallandier, 2006.

Deutsch, Karl, *Nationalism and Social Communication: An Inquiry into the Foundation of Nationality,* Cambridge, MA: MIT Press, 1966.

De Waal, Alex, "Creating Devastation and Calling it Islam," *SAIS Review,* 21.1 (2001): 117–32.

Diamond, Jared, *Guns, Germs, and Steel: The Fates of Human Societies,* New York: Norton, 1999.

Diamond, Larry, *Developing Democracy,* Baltimore, NJ: Johns Hopkins University Press, 1999.

Doornbos, Martin, "State Collapse, Civil Conflict and External Intervention," in Peter Burnell and Vicky Randall, *Politics in the Developing World, 2nd edn.,* Oxford: Oxford University Press, 2008: 249–67.

Dos Santos, Theotino, "The Structure of Dependence," *American Economic Review,* 60 (May 1970).

Drescher, S., and S. Engerman, eds, *A Historical Guide to World Slavery,* Oxford: Oxford University Press, 1998.

Duverger, Maurice, *Political Parties,* New York: Wiley, 1954.

Edison, H., "Testing the Links: How Strong Are the Links between Institutional Quality and Economic Performance?" *Finance and Development,* Vol. 40, No. 2, 2002: 35–7.

EIA, *Short-term Energy Outlook, Country Analysis Brief,* October 1007, http://www.eia.doe.gov/emeu/cabs/Equatorial Guinea/Full.html.

EITI, "Extractive Industries Transparency Initiative Sourcebook," available on-line from www.eitransparency.org/document/sourcebook.

El Mundo (December 14, 2003) "El 'crudo' Panorama de la Guinea de Obiang".

Eltigani, Ateem, "Anatomy of the Conflict in Darfur," in Deng (2009): 253–74.

Evans, Peter, *Dependent Development: The Alliance of Multinational, State, and Local Capital in Brazil*, Princeton, NJ: Princeton University Press, 1979.

ExxonMobil, "Tapping into a New Frontier Oil Province," October 1998, cf. Gary and Karl 2003: 60, www.esso.com/eqff/essochad/news/press_oct98/main.html.

Finer, S.E., *The Man on Horseback: The Role of the Military in Politics*, London: Pall Mall Press, 1962.

First, Ruth, "Libya, Class and State in an Oil Economy," in Peter Nore and Terisa Turner, eds, *Oil and Class Struggle*, London: Zed Books, 1980.

Foltz, William, "Reconstructing the State in Chad," in Zartman 1995: 15–32.

Frank, André Gunder, *Dependent Accumulation and Underdevelopment*, New York: Monthly Review Press, 1979.

Frank, Isaiah, *Foreign Enterprise in Developing Countries: A Supplementary Paper of the Committee for Economic Development*, Baltimore, NJ: Johns Hopkins University Press, 1980.

Frynas, George, George Wood, and Ricardo Soares de Oliveira, "Business and Politics in São Tomé e Príncipe: From Cocoa Monoculture to Petrol-State," *African Affairs*, No. 102: 51–80.

Gardinier, David, and Douglas Yates, *The Historical Dictionary of Gabon, 3rd edn*, Metuchen, NJ: Scarecrow Press, 2006.

Gary, Ian, and Karl, Terry Lynn, *Bottom of the Barrel: Africa's Oil Boom and the Poor*, Baltimore, MD: Catholic Relief Services, 2003.

Gelb, Alan, and Associates, *Oil Windfalls: Blessing or Curse?* World Bank Research Publication, Oxford: Oxford University Press, 1988.

Ghazvinian, John, *Untapped: The Scramble for Africa's Oil*, Orlando, FL: Harcourt, 2007.

Gibbins, Roger, *Prairie Politics and Society: Regionalism in Decline*, Toronto: Butterworths, 1980.

— *Regionalism: Territorial Politics in Canada and the United States*, Toronto: Butterworths, 1982.

Glaser, Antoine, and Stephen Smith, *Comment la France a Perdu l'Afrique*, Paris: Calmann-Levy, 2004.

Global Witness, *A Crude Awakening: The Role of the Oil and Banking Industries in Angola's Civil War and the Plunder of State Assets*, December 1999, www.globalwitness.org.

— *Time for Transparency: Coming Clean on Oil, Mining and Gas Revenues*, March 2004, www.globalwitness.org.

— *The Secret Life of a Shopaholic: How an African Dictator's Playboy Son Went on a Multi-Million Dollar Shopping Spree in the US* (November 2009).

— *Undue Diligence: How Banks Do Business with Corrupt Regimes* (March 2009).

Goodstein, David, *Out of Gas: The End of the Age of Oil*, New York: Norton, 2004.

Gott, Richard, *In the Shadow of the Liberator: Hugo Chávez and the Transformation of Venezuela*, London: Verso, 2005.

Greenpeace, "The Bonnie and Clyde of Karmabanque" (June 4, 2003), www.greenpeace.org.

Hartz, Louis, *The Founding of New Societies*, New York: Harcourt, Brace & World, 1964.

Hellman, J.S., G. Jones, and D. Kaufman, "'Seize the State, Seize the Day,' State Capture, Corruption and Influence in Transition," World Bank Policy Research Working Paper N° 2444, Washington. DC: World Bank Institute, 2000.

Herbert, Bob, "A Radical Treasure," *New York Times* (January 30, 2010): A17.

Heywood, Paul, ed., *Political Corruption*, Oxford: Blackwell, 1997.

Higgot, R.A., and M. Ougaard, eds, *Towards a Global Polity*, London: Routledge, 2002.

Higman, Barry, "Demography," in Drescher and Engerman 1998.

Hodges, Tony, *Angola: From Afro-Stalinism to Petro-Diamond Capitalism*, Oxford: James Currey, 2001.

Homer-Dixon, Thomas, *Environment, Scarcity and Violence*, Princeton, NJ: Princeton University Press, 2000.

Howarth, Stephen, Joost Jonker, Keetie Sluyterman, and Jan Luiten van Zanden, *The History of the Royal Dutch Shell*, Oxford: Oxford University Press, 2007.

Human Rights Watch, "Letter to President Obasanjo" (April 4, 2003), www.hrw.org/press/2003/04/nigeria0400703obasanjo.htm.

Hutchcroft, Paul, "The Politics of Privilege: Assessing the Impacts of Rents, Corruption, and Clientism on Third World Development," in Heywood 1997: 223–42.

Huntington, Samuel, *Political Order in Changing Societies*, New Haven, CT: Yale University Press, 1968.

— *The Third Wave: Democratization in the Late Twentieth Century*, Norman: University of Oklahoma Press, 1991.

Idris, Amir, *Sudan's Civil War: Slavery, Race and Formational Identities*, Lewiston, NY: Edwin Mellon Press, 2001.

IMF, Public Information Notice N° 03/144 (December 9, 2003).

Janowitz, Morris, *The Military in the Political Development of New Nations*, Chicago, IL: University of Chicago Press, 1964.

Johnson, Douglas, *The Root Causes of Sudan's Civil Wars*, Oxford: James Currey, 2003.

Kaplan, Robert, *The Coming Anarchy: Shattering the Dreams of the Post Cold War*, New York: Vintage Books, 2000.

Karl, Terry Lynn, *The Paradox of Plenty: Oil Boom and Petro-States*, Berkeley: University of California Press, 1997.

Kaufmann D., A Kraay, and M. Mastruzzi, "Governance Matters VII: Aggregate and Individual Governance Indicators for 1996–2007," World Bank Policy Research Working Paper N° 4654, Washington, DC: World Bank Institute, 2008.

Khalid, Mansour, ed., *John Garang Speaks*, London/New York: Routledge & Kegan Paul, 1987.

Klare, Michael, *Resource Wars: The New Landscape of Global Conflict*, New York: Henry Holt, 2001.

Klitgaard, Robert, *Controlling Corruption*, Berkeley: University of California Press, 1988.

— *Tropical Gangsters: One Man's Experience with Development and Decadence in Deepest Africa*, New York: Basic Books, 1990.

Krueger, Anne O., "The Political Economy of the Rent-Seeking Society," *American Economic Review*, 64 (1974): 291–303.

Laakso M., and R. Taagepera, "Effective Number of Parties: A Measure with Applications to West Europe," *Comparative Political Studies*, 12 (1979): 3–27.

Le Pere, Garth, ed., *China in Africa: Mercantilist Predator, or Partner in Development?* Johannesburg: Institute for Global Dialogue/South African Institute of International Affairs, 2006.

Lebi, Simplice Euloge, *Pour une Histoire militaire du Congo-Brazzaville 1882–1992: Problèmes et perspectives de l'administration militaire,* Paris: l'Harmattan, 2009.

Leftwich, A. "Bringing Politics Back In: Towards a Model of the Developmental State," *Journal of Development Studies,* Vol. 31, No. 3 (1995): 400–27.

Lei, Wu, "Le pétrole, la question du Darfur et le dilemma chinois," *Outre-Terre: Revue Française de Géopolitique* N° 20 (2008): 215–26.

Lenin, Vladimir, *The Essential Lenin,* New York: Bantam Books 1966 [1916].

Liniger-Goumaz, Max, *Small Is Not Always Beautiful: The Story of Equatorial Guinea,* London: Hurst & Co., 1989.

— *United States, France and Equatorial Guinea: The Dubious "Friendship"*,Geneva: Les Editions du Temps, 1997.

— *Historical Dictionary of Equatorial Guinea, 3rd edn,* Lanham, MD: Scarecrow Press, 2000.

— *Colonisation—Néocolonialisation—Démocratisation— Corruption: L'Aune de la Guinée Equatoriale,* Geneva: Editions du Temps, 2003.

— *La Guinée Equatoriale Convoitée et Opprimée: Aide-mémoire d'une démocrature 1968–2005,* Paris: L'Harmattan, 2005.

Locke, John, "On Property," *Second Treatise of Government,* Indianapolis: Hackett Publishing Company, 1980 [1690].

McAdam, Douglas, *Political Process and the Development of Black Insurgency, 1930–1970,* Chicago, IL: University of Chicago Press, 1982.

M'Bokolo, Elikia, *L'Afrique Noire: Histoire et civilisations, Tome II, XIXeme–XXeme siècles,* Paris: Hatier/AUPELF, 1992.

Mahdavy, Hossein, "Patterns and Problems of Economic Development in Rentier States: The Case of Iran," in M.A. Cook, ed., *Studies in the Economic History of the Middle East,* Oxford: Oxford University Press, 1970.

Malthus, Thomas, *An Inquiry into the Nature and Progress of Rent,* London: John Murray 1815.

Manning, Patrick, *Slavery and African Life: Occidental, Oriental and African Slave Trades,* Cambridge: Cambridge University Press, 1990.

Martin, Guy, *Africa in World Politics: A Pan-African Perspective,* Trenton, NJ/Asmara: Africa World Press, 2002.

Médard, Jean-François, "Le Big Man en Afrique : Esquisse d'Analyse du Politicien Entrepreneur," *L'Année Sociologique,* No. 42 (1992).

Menard, Scott, *Longitudinal Research, 2 2nd edn,* London: Sage, 2002.

Miller, Joseph C. *Way of Death: Merchant Capitalism and the Angolan Slave Trade, 1730-1830,* Madison: University of Wisconsin Press, 1988.

Minter, William, *Portuguese Africa and the West,* New York: Monthly Review Press, 1972.

Moore, Mick "Revenues, State Formation, and the Quality of Governance in Developing Countries," *International Political Science Review,* 25: 3 (2004) 297–319.

Morin-Allory, Roman, "Chine–Sudan, une amitié à l'ombre des derricks," *Outre-Terre: Revue Française de Géopolitique,* N° 20 (2008): 227–43.

Naanen, B., "Oil-producing Minorities and Restructuring of Nigerian Federalism: The Case of the Ogoni Peoples," *Journal of Commonwealth and Comparative Politics,* Vol. 33, No. 1 (1995): 47–8.

Nairn, T. ,*The Break-up of Britain: Crisis and Neo-Nationalism*, London: New Left Books, 1977.

Nkrumah, Kwame, *Class Struggle in Africa*, New York: International Publishers, 1970.

— *Towards Colonial Freedom: Africa in the Struggle Against World Imperialism*, London: Zed Books, 1973.

Omeje, Kenneth, *High Stakes and Stakeholders: Oil Conflict and Security in Nigeria*, Aldershot: Ashgate, 2006.

— "Re-Engaging Rentier Theory and Politics," in Kenneth Omeje, ed., *Extractive Economies and Conflicts in the Global South: Multi-Regional Perspectives on Rentier Politics*, Hampshire: Ashgate, 2008: 1–25.

— *Extractive Economies and Conflicts in the Global South: Multi-Regional Perspectives on Rentier Politics*, Aldershot: Ashgate, 2008.

Onoh, J.K., *The Nigerian Oil Economy*, New York: St. Martin's Press, 1983.

Open Budget Initiative, www.openbudgetindex.org.

Osaghae, E., Onwudiwe, E., and R. Suberu, eds, *The Nigerian Civil War and its Aftermath*, Ibadan: John Archers, 2002.

Owen, Robert, and Bob Sutcliffe, *Studies in the Theory of Imperialism*, 7th edn, Harlow, Essex: Longman 1981 [1972].

Oyono Sa Abegue, V., *L'Evolution des structures productives et socials de l'économie de la Guinée Equatoriale (1858–1968)* Doctoral Thesis, University of Lyon, 1985.

Péan, Pierre, *Affaires Africaines*, Paris: Fayard, 1983.

Pierce, William Spangar, *Economics of the Energy Industries*, Belmont, CA: Wadsworth, 1986.

Plato, *The Republic, Books VI–X*, trans. Paul Shorey, Loeb Classical Library, Cambridge, MA: Harvard University Press, 1987.

Prebisch, Raul, *The Economic Development of Latin America and Its Principal Problems*, New York: United Nations, 1950.

Prunier, Gérard, *Africa's World War: Congo, the Rwandan Genocide, and the Making of a Continental Catastrophe*, Oxford: Oxford University Press, 2009.

Reno, William, *Corruption and Politics in Sierra Leone*, Cambridge: Cambridge University Press, 1995.

Ricardo, David, *The Principles of Economy and Taxation*, London: Everyman's Library, 1821.

Robinson, Ronald, "Non-European Foundations of European Imperialism: Sketch for a Theory of Collaboration," in Owen and Sutcliffe 1972: pp. 117–40.

Rodney, Walter, *How Europe Underdeveloped Africa*, Washington, DC: Howard University Press, 1974.

Ross, Michael, "The Natural Resource Curse: How Wealth Can Make You Poor," in Ian Bannon and Paul Collier, eds, *Natural Resources and Violent Conflict: Options and Actions*, Washington, DC: World Bank, 2003.

Rotberg, Robert, *When States Fail: Causes and Consequences*, Princeton, NJ: Princeton University Press, 2004.

Rothchild, Donald, "State–Ethnic Relations in Middle Africa," in Gwendolen Carter and Patrick O'Meara, eds, *African Independence: The First Twenty-Five Years*, Bloomington: Indiana University Press, 1985.

Sala-i-Martin, Xavier, and Arvind Subramanian, "Addressing the Natural Resource Curse: An Illustration from Nigeria," IMF Working Paper (July 2003).

Salih, Mohamed, "Kordofan: Between Old and New Sudan," in Deng 2009: 275–306.

Same, Achille Toto, *Mineral-Rich Countries and Dutch Disease: Understanding the Macroeconomic Implications of Windfalls and the Development Prospects: The Case of Equatorial Guinea*, Policy Research Working Paper 4595, Washington, DC: World Bank, April 2008.

Sampson, Anthony, *The Seven Sisters: The Great Oil Companies and the World They Shaped*, New York: Viking Press, 1975.

Sassou-Nguesso, Denis, *Straight Speaking for Africa: Interviewed by Edouard Dor*, Trenton, NJ/Asmara: Africa World Press, 2009.

Save the Children, *Lifting the Resource Curse: Extractive Industry, Children and Governance*, London: Save the Children, 2003.

Schumpeter, Joseph, *Capitalism, Socialism, and Democracy*, New York: Harper & Brothers, 1947.

Seibert, Gerhard, "São Tomé e *Príncipe*," in Andreas Mehler, Henning Melber and Klaas van Walraven, eds, *Africa Yearbook: Politics, Economy and Society South of the Sahara in 2003*, Leiden: Brill, 2004.

Seibert, Gerhard, "São Tomé e Príncipe: The Difficult Transition from International Aid Recipient to Oil-Producer," in Matthias Basedau and Andreas Mehler, eds, *Resource Politics in Sub-Saharan Africa*, Hamburg: Institut für Afrika-Kunde, 2005: 223–50.

— *Comrades, Clients and Cousins: Colonialism, Socialism and Democratization in São Tomé e Príncipe*, Leiden: Brill, 2006.

Shaxson, Nicholas, *Poisoned Wells: The Dirty Politics of African Oil*, New York: Palgrave MacMillan, 2008.

Sen, Amartya, *Identity and Violence: The Illusion of Destiny*, London: Penguin, 2007.

Shils, E. *Center and Periphery: Essays in Macrosociology*, Chicago, IL: Chicago University Press, 1975.

Silverstein, Ken, "Oil Boom Enriches African Ruler," *Los Angeles Times* (January 20, 2003).

Soares de Oliveira, Ricardo, *Oil and Politics in the Gulf of Guinea*, London: Hurst, 2007.

Somerville, Keith, *Angola*, New York: St. Martin's Press, 1986.

Stockwell, John, *In Search of Enemies: A CIA Story*, New York: W.W. Norton, 1984.

Takougang, Joseph, "Cameroon: Biya and Incremental Reform," in John F. Clark and David Gardinier, eds, *Political Reform in Francophone Africa*, Boulder, CO: Westview Press, 1997: 162–81.

Thomson, Alex, *An Introduction to African Politics*, 2nd edn, New York: Routledge, 2004.

Thucydides, *History of the Peloponnesian War, Books V–VI*, trans. C.F. Smith, Cambridge, MA: Harvard University Press , 1977 [1921].

Traub-Merz, Rudolf, and Douglas Yates, eds, *Oil Policy in the Gulf of Guinea: Security and Conflict, Economic Growth, Social Development*, Bonn: Friedrich Ebert Stiftung, 2004.

Transparency International, 2007 Corruption Perception Index <www.transparency.org> (2008).

Turner, T. "Nigeria: Imperialism, Oil Technology and the Comprador State," in P. Norre and T. Turner, eds, *Oil and Class Struggle*, London: Zed Books, 1980: 199–223.

Ukiwo, Ukoha, "Nationalization versus Indigenization of the Rentier Space: Oil and Conflicts in Nigeria," in Omeje, ed., 2008: 75–92.

Wallerstein, Immanuel, *The Modern World System*, New York: Humanities Press, 1974.

Wall Street Journal, "China Discusses Darfur Oil-Hunt Aid," (July 9, 2008).

Watts, Michel, "Anatomy of an Oil Insurgency: Violence and Militants in the Niger Delta, Nigeria," in Omeje 2008: 51–74.

Weber, Max, *Economy and Society: An Outline of Interpretive Sociology*, Berkeley: University of California Press, 1978 [1922].

Weblen, Thorstein, *The Theory of the Leisure Class*, intro. John Kenneth Galbraith, Boston: Houghton Mifflin 1973 [1899].

Willetts, Peter, *The Non-Aligned Movement: The Origins of a Third-World Alliance*, London: Frances Pinter, 1978.

Woo-Cummings, M., ed., *The Developmental State*, Ithaca, NY: Cornell University Press, 1999.

World Bank, *Equatorial Guinea: An Introductory Economic Report*, June 1983.

— *Technical Assistance Project—Republic of Equatorial Guinea*, May 1984.

— *Equatorial Guinea*, Report No. P-4288, April 1986.

— *Project Appraisal Document, Chad/Cameroon Petroleum, Petroleum Development and Pipeline Project*, April 13, 2000.

— "Managing Oil Revenues in Chad: Legal Deficiencies and Institutional Weaknesses," Harvard Law School, October 13, 1999, cf. Gary and Karl 2003: 73.

— *Project Appraisal Document (PAD), Chad/Cameroon Petroleum Development and Pipeline Project*, April 13, 2000, cf. Gary and Karl 2003: 24.

— *African Economic and Financial Data*, Washington, DC: World Bank, 1986.

— *2007 World Governance Indicators* <www.worldbank.org>, 2008.

Yama Nkounga, Albert, "Les Fonds Petroliers," in Traub-Merz and Yates 2004: 183–8.

Yates, Douglas, *The Rentier State in Africa: Oil-Rent Dependency and Neocolonialism in the Republic of Gabon*, Trenton, NJ/Asmara: Africa World Press, 1996.

— "The Scramble for African Oil," *South African Journal of International Affairs*, Vol. 13, No. 2 (Winter/Spring 2006): 11–31.

— "Gabon," in Andreas Mehler, Henning Melber, and Klaas van Walraven, eds, *Africa Yearbook: Politics, Economy and Society South of the Sahara in 2007*, Leiden/Boston: Brill, 2008.

— *The French Oil Industry and the Corps des Mines in Africa*, Trenton, NJ/Asmara: Africa World Press, 2009.

Yergin, Daniel, *The Prize: The Epic Quest for Oil, Money, and Power*, New York: Simon & Schuster, 1991.

Young, Crawford, "Evolving Modes of Consciousness and Ideology: Nationalism and Ethnicity," in David Apter and Carl Rosberg, eds, *Political Development and the New Realism in Sub-Saharan Africa*, Charlottesville, University Press of Virginia, 1994.

Zakaria, Fareed, "The Rise of Illiberal Democracy," *Foreign Affairs*, Vol. 76, No. 6 (1997): 22–43.

Zartman, William, *Collapsed States: The Disintegration and Restoration of Legitimate Authority*, Boulder, CO: Lynne Reinner, 1995.

Zolberg, Aristide, *Creating Political Order: The Party States of West Africa*, Chicago, IL: Rand McNally, 1966.

Index

Compiled by Sue Carlton